INTERNATIONAL SPACE STATION

1998–2011 (all stages)

To Ann, the love of my life, soul-mate and fellow 'philosopher' from the school of life and the university of hard knocks. For everything we are to each other.

First published in October 2012

A catalogue record for this book is available from the British Library

ISBN 978 085733 218 9

Library of Congress control no. 2012938591

Published by Haynes Publishing,
Sparkford, Yeovil,
Somerset BA22 7JJ, UK.
Tel: 01963 442030 Fax: 01963 440001
Int. tel: +44 1963 442030 Int. fax: +44 1963 440001
E-mail: sales@haynes.co.uk
Website: www.haynes.co.uk

Haynes North America Inc.
861 Lawrence Drive, Newbury Park,
California 91320, USA.

Printed in the USA by Odcombe Press LP,
1299 Bridgestone Parkway, La Vergne, TN 37086.

Acknowledgements

The author and publisher would like to thank NASA, the European Space Agency, the Russian Space Agency, the Japan Aerospace Exploration Agency, the Canadian Space Agency and a wide range of commercial and aerospace manufacturing companies for providing information and images for this book.

Useful contacts

British Interplanetary Society
27/29 South Lambeth Road, London SW8 1SZ, UK
Tel 020 7735 3160
www.bis-spaceflight.com
Signatory to the International Astronautical Federation, the BIS was formed in 1933 with Arthur C. Clarke an early member. The Society is open to all and has two publications available on subscription. Holds regular meetings with lectures and has a library open to members.

National Space Centre
Exploration Drive, Leicester LE4 5NS, UK
Tel 0845 605 2001
www.spacecenter.co.uk
Provides an all-round educational experience in many separate aspects of space research and exploration.

The Science Museum
Exhibition Rd, South Kensington SW7 2DD, UK
Tel 0870 870 4868
www.sciencemuseum.org.uk
Contains a full space gallery with many relevant exhibits including a Shuttle model and artifacts from the space programmes of Europe, the UK, the US, Russia and China.

International Space University
Parc d'Innovation, 1 rue Jean-Dominique Cassini, 67400 Illkirch-Graffenstaden, France
Tel 0033 3 88 65 54 30
www.isunet.edu
Provides graduate-level educational resources, courses and qualifications for the aspiring space professional, having graduated more than 3,000 students from 100 countries.

Kennedy Space Center Visitor Complex
SR405 Kennedy Space Center, Florida 32899, USA
Tel 001 866 737 5235
www.kennedyspacecenter.com
Situated close to the Kennedy Space Center from where NASA launched all its 166 manned space missions, this is a large parkland of exhibits, displays, rockets, Shuttle mock-up to enter, exhibits, displays and guided tours. IMAX theatre and everyday outdoor informal lunch with an astronaut.

National Air & Space Museum
6th & Independence Avenue SW, Washington DC 20560, USA
Tel 001 202 633 1000
www.nasm.si.edu
The world's largest aerospace museum with numerous galleries and exhibits covering the space programme and the Shuttle specifically.

Space Center Houston
1601 NASA Parkway, Houston, Texas 77058, USA
Tel 001 281 244 2100
www.spacecenter.org
Contains one of the world's largest collections of space exhibits with tours of Shuttle-related facilities right alongside the NASA Johnson Space Center.

Steven F Udvar Haizy Center
14390 Air & Space Museum Parkway, Chantilly, VA 20151, USA
Tel 001 202 633 1000
www.nasm.si.edu/udvarhaizy
Aerospace museum with the Shuttle Orbiter OV-101 Enterprise as the lead display.

US Space and Rocket Center
One Tranquility Base, Huntsville, Alabama 35805, USA
www.spacecamp.com/museum
Contains one of the largest displays of Shuttle-related hardware and exhibits anywhere and provides educational opportunities with IMAX theatre.

COVER CUTAWAY: *John Lawson*

INTERNATIONAL SPACE STATION

1998–2011 (all stages)

Owners' Workshop Manual

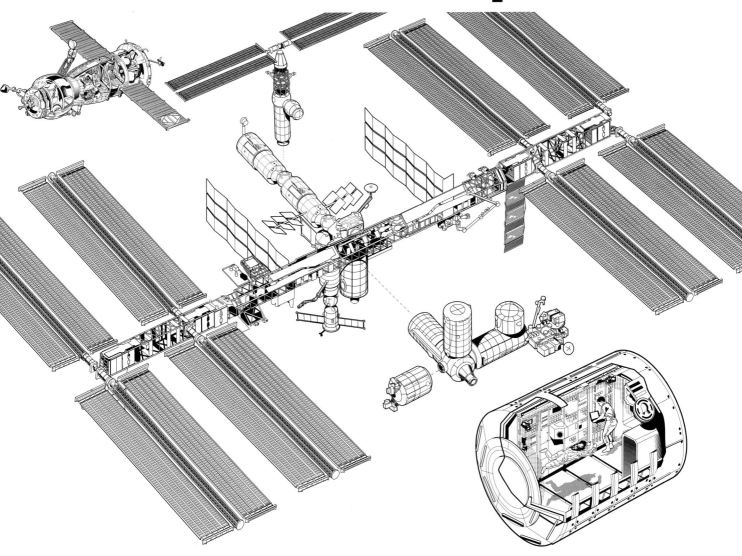

An insight into the history, development, collaboration, production and role of the permanently manned earth-orbiting complex

David Baker

Contents

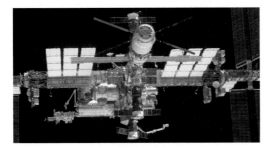

OPPOSITE Set against an ice-clad portion of the Earth's surface, veiled only sightly by clouds more than 200 miles below, the completed International Space Station is a majestic tribute to the work of 15 countries over several decades. *(NASA)*

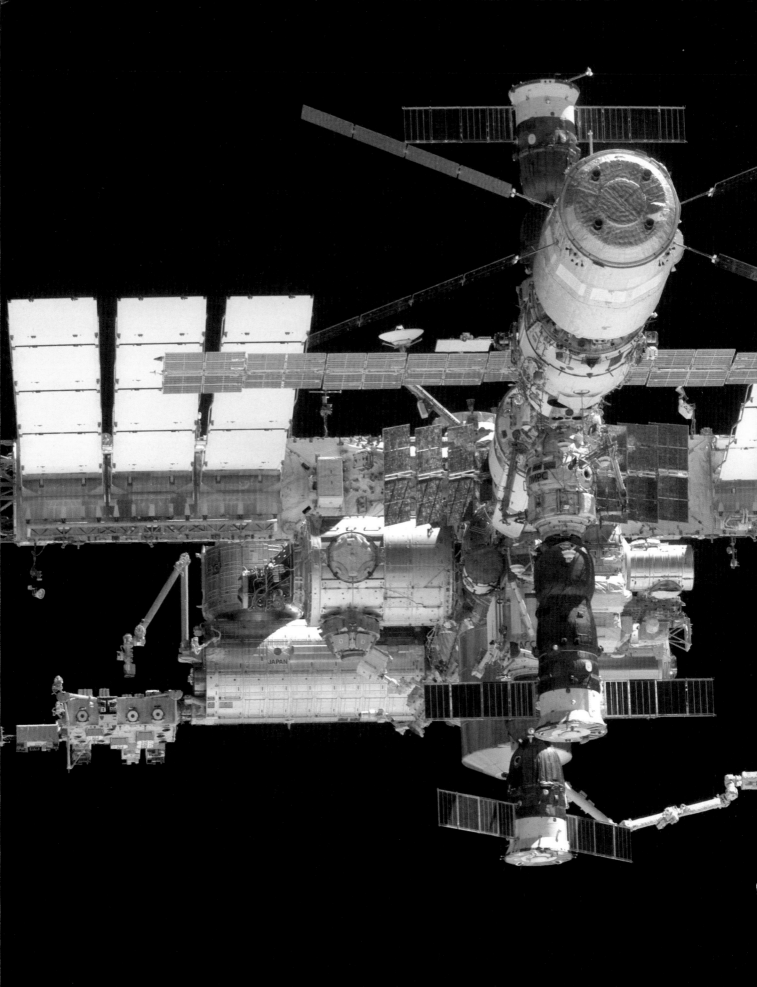

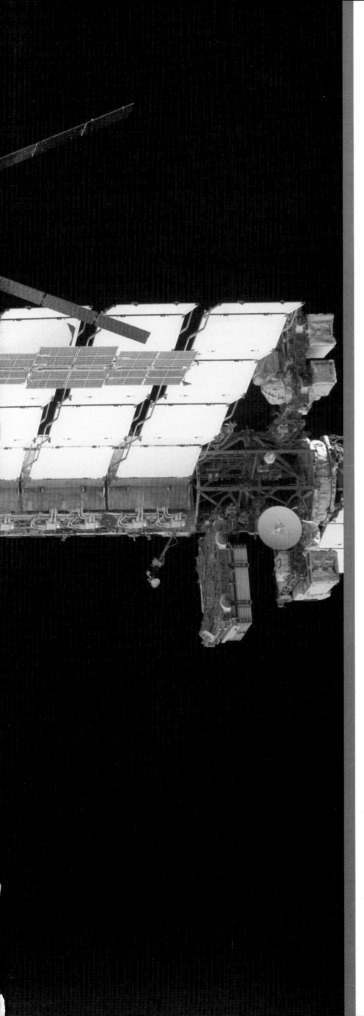

Introduction

The International Space Station (ISS) has been long in coming and has taken more than 12 years to build, but it has united 15 countries and five space agencies in an unprecedented effort to construct a giant orbiting facility that may outlast many who helped design it.

OPPOSITE With Russian Soyuz manned taxi spacecraft and Progress cargo-tankers top and bottom and Europe's Automated Transfer Vehicle in the foreground, the International Space Station is a truly global science platform. *(ESA)*

ABOVE The Shuttle STS-126 crew (red T-shirts) poses with the six-member ISS crew. *(NASA)*

BELOW Canada's robotic arm projects from the ISS over a sun-drenched planet below. *(CSA)*

Forged from the tools of a Cold War, the ISS is a testament to cooperation, not only between amicable partners but also between former adversaries, bridging the ideological divide that once seemed permanent. Historians of both politics and technology assert that herein is its true value. Where once it was said that after landing on the Moon, for mankind only the limitations of imagination acts as a constraint on the possible, now it can be said that if the world can unite behind this great international venture, surely a world divided by differences must become a thing of the past.

Yet the ISS is more than a complex assembly of modules, truss structures and solar wings. It is the culmination of efforts in many countries for humans to remain in space for long periods, experience that dates back to the early 1970s when both Russia and the USA launched the first habitats capable of supporting people for long periods in the challenging environment of weightlessness. The ISS is a direct product of those early attempts to build a human presence in space and as such it builds on early efforts without which the success of the International Space Station would be impossible. The Russians were first with their Salyut space station, launched in 1971, followed by the Americans with the Skylab station just two years later. A major participant in robotics on the ISS, Canada contributed first by building the Shuttle's remote manipulator arm, first flown in 1981 on the second Shuttle flight. And European expertise was built upon participation when Germany led a consortium of nations to build the first orbital laboratory called Spacelab, first carried by the Shuttle in 1983.

Today, the ISS is complete in its definitive design configuration but it will continue to change and adapt to the needs of scientists and research workers around the globe who routinely visit the station, use its tools and profit from the results of myriads of experiments inside and outside this 400-ton (363,000kg) complex. It is assured a further life until at least 2020 but participating agencies are already talking about maintaining the station until 2028 or later, when new generations of spacecraft may use it as a springboard to the deeper regions of the solar system. From this facility research will benefit people on Earth and provide information for engineers planning mankind's first forays beyond the Earth–Moon system, perhaps as far as Mars and the asteroids. New processes for medicines, materials and biological agents are even now being tested and from the ISS we may learn more about the human body and its ageing process than we ever could by remaining on Earth.

Uniting nations in a cooperative venture where interdependence replaces isolation, where convergence replaces conflict, and where future aspirations are shared and not opposed, allows a new generation of young idealists to work in a truly global village to suppress the petty hatreds of bitterness and rivalry that once fuelled the nascent space programme, where competition

AN IDEAL HOME IN SPACE

In one of the more unlikely steps to building a space station, an exposition in London, England, known as the Ideal Home Exhibition, would be the venue for the first public display of the way NASA would get its first station in space. With space on the world agenda and public interest at a high, the UK's *Daily Mail* newspaper sent a reporter to visit US aerospace companies and find out what they thought about future possibilities of living in space and what such stations would look like. He visited the Douglas Aircraft Company in Santa Monica, California, who were only too pleased to publicise their work on such an application.

Several noted rocket engineers, including von Braun, had proposed the use of spent rocket stages as an economical way of putting a station into space. Retain the last stage of a Saturn rocket, vent the residue of its remaining fuel into the vacuum of space and convert the cavernous interior of the propellant tank into a habitable place to work. Pressurised with oxygen and nitrogen, it could be made to support teams of astronauts ferried back and forth by a multi-man successor to Mercury or a space-plane like the Air Force Dyna-Soar.

Douglas liked that idea and, inspired by

LEFT NASA's Rene Berglund in 1963 with a model of a space station concept comprising linked modules launched by a giant rocket (foreground). *(NASA)*

the nudge from one of the UK's leading newspapers, a team was set to work under W. Nissam with a budget of $10,000. The company already had the contract to build the S-IV second stage for Saturn and in their concept adapted that as a potential habitat in space. Four petal-like doors covering the forward section which supported the manned spacecraft for launch would open in orbit, revealing solar cells for electrical power.

The newspaper liked the idea and asked Douglas to build a full-scale replica of the concept for public display at the *Daily Mail* Ideal Home Exhibition in March 1960, a full-size structure through which members of the public could walk – and 200,000 did just that, getting their first experience of what a space station might look like!

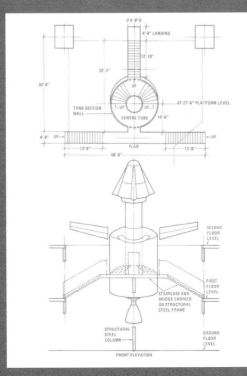

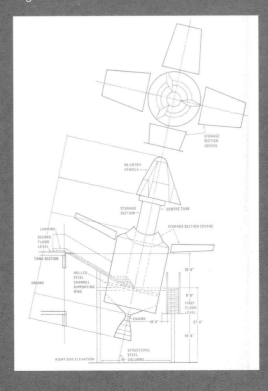

FAR LEFT The *Daily Mail* inspired a space station concept put together by the Douglas Aircraft Company for the London Ideal Home Exhibition. *(David Baker)*

LEFT The Douglas space station concept was set up as a full-scale, walk-through, replica. *(David Baker)*

is a proud thing and not a springboard to violence. As cargo ships rise from launch pads in Japan and from the coast of South America, as freighters lift off from the sprawling complex at Baikonur and as commercial companies send logistical supplies to the ISS from Cape Canaveral, those who foresaw an orbiting space station as an opportunity for mankind to work together in peace and harmony will be pleased with what they started.

Stepping stones to a habitat

The tortuous path that NASA (National Aeronautics and Space Administration) would follow to build a space station in orbit was laid in politically motivated decisions set by presidents and lawmakers. Not for 40 years after it was formed would NASA make a start on its prime goal – constructing a permanent workplace in space. When it came it was very different from the original concept developed from ideas evolved over the preceding century and it would only happen at all because of the collapse of the Soviet Union.

From a Power Tower to Freedom

Tasked with sending men to the Moon by the end of the 1960s, NASA grew far beyond the expectations of its founders. It planned ambitious ventures to far-flung

destinations but stuck to an age-old concept of the space station as a springboard to the planets. But in achieving the Moon goal, NASA saw those dreams collapse and searched in vain for the way to build a habitat in space.

On 25 January 1984, at the annual State of the Union address to a joint session of Congress, President Reagan authorised NASA to begin work on a space station 'and to do it within a decade'. In February, NASA administrator Jim Beggs set out to tour the pro-Western world and conduct presentations to invite interest beginning with a visit to the UK, where he was given a cool reception by the Thatcher government. From there he visited West Germany, Italy, and France before flying to Japan. Continental European countries gave Beggs a warm reception and West Germany in particular was keen to participate in the space station as part of the European Space Agency (ESA). Japan too was especially excited about the prospect of making further progress in its national space effort and, having already designed and produced the Shuttle's robotic arm, Canada was eager to develop advanced robotics for the station.

The configuration incorporated three elements: a central core structure of four laboratory and habitation modules supporting a crew of eight in a 28.5-degree orbit; an unmanned co-orbiting platform for free-flyer experiments; and a second unmanned free-flyer orbiting the Earth at 90 degrees to the equator. But as Jim Beggs gathered his list of partners, the design was already in a state of flux. Throughout the summer three design options were considered: the 'Power Tower', the 'Planar' station and an odd configuration known as the 'Delta' station. Originated by McDonnell Douglas and Grumman, the Power Tower consisted of a 300ft (91m) tall lattice tower across which would be attached a 200ft (61m) wide cross-beam mounted two-thirds of the way up, with four Solar Array Wings at each end. Up to five modules, each up to 35ft (10.7m) long, could be attached to the bottom of the tower with large radiator wing panels on either side. This configuration would always fly with the modules pointing Earthwards and achieve some stability from a pendulous alignment with the Earth's gravity. The Planar station concept consisted of a single rigid truss assembly 300ft (91m) long with the modules attached at the

BELOW Boeing's Space Operations Center envisaged modules for research and satellite servicing. *(Boeing)*

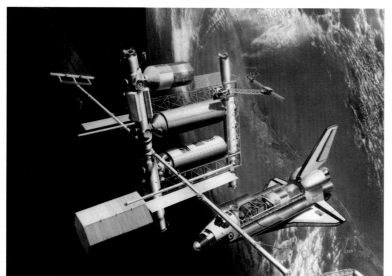

centre and four large Solar Array Wings at each end. Science experiments would be attached to a large A-frame extending 80ft (24m) above the truss. The Delta station derived its name from an inverted triangular shape at the top of which would be a 175ft by 125ft (53m by 38m) platform of solar cells covering 28,600ft^2 (2,657m^2). Extending down from this platform would be an inverted V-shaped field of beams supporting five pressurised modules at the bottom.

Within months the Power Tower became the reference baseline. Europe proposed a pressurised experiment module it would call 'Columbus', lifted to the station by Shuttle where it would form one of the cluster of modules, including those from the US and one from Japan. By mid-1985 the station configuration had shifted from the Power Tower concept to a dual-keel design, a configuration increasingly favoured since the beginning of the year. This envisaged a transverse truss holding solar arrays midway across a rectangular lattice structure. The total span of the solar array truss would now be 503ft (153m), supporting at its centre the pressurised living and work modules.

There would be only two US modules, each increased in length from 35ft to 44.5ft (10.7m to 13.4m) but retaining a diameter of 13.6ft (4.15m), attached in proximity to the modules from Europe and Japan. The parallel long sides of the rectangle supported by the truss would each be about 330ft (101m) tall, 126ft (38m) apart and connected at the bottom and at the top. Overall height would be 361ft (110m). The top cross-section would overhang the vertical sides to provide a total span of 297ft (91m), providing space for instruments and equipment. The upper cross-section would be used for astrophysical observations and the lower one for Earth observations with the idea being to 'fly' the station perpendicular to the path of the orbit.

The dual-keel design would have only four solar array wings, adding dish-shaped solar heat collectors to drive alternators producing 400-cycle AC electric power. This reduced the large surface area of the station, causing drag from the random air molecules present even in low Earth

RIGHT Large payloads could be placed around the rectangular-shaped truss structure for ease of access by the Shuttle. *(NASA)*

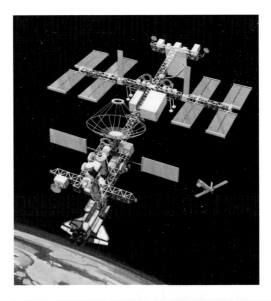

LEFT In 1984 NASA got formal approval to begin development of a space station, this concept known as the Power Tower put all the solar arrays at one end of the assembly. *(NASA)*

BELOW By 1986 the design had shifted to the Dual Keel concept but retained provision for satellite servicing. *(NASA)*

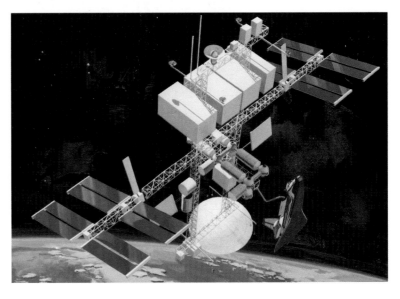

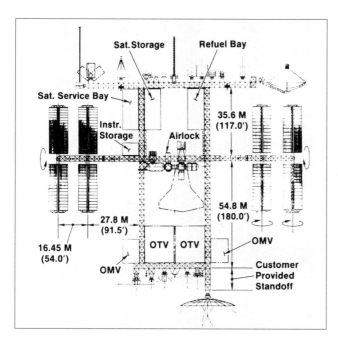

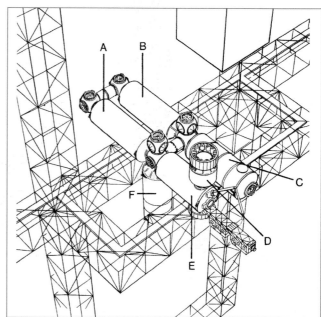

orbit. The problem with the Power Tower had been instability. This was avoided by the dual-keel design, which, with the modules located at the centre, had better balance. Europe got to work designing its Columbus module while Japan began a detailed design of its own module.

When approved by Reagan in 1984 it had been hoped to build the station with 8–10 flights, but that estimate had now increased to at least 19, and possibly as many as 31 flights. Questions also emerged regarding the amount of space-walking, or EVA (extra-vehicular

activity), necessary to build and maintain the dual-keel station and over the need for a lifeboat rescue vehicle. As 1989 drew to a close it looked as though the station, named 'Freedom', may be on track for a first element launch in March 1995. But then a special review group concluded there was an impossibly high reliance on EVA, which is inherently dangerous and saps valuable work time otherwise spent on science experiments. Just to keep the station going, said the group, astronauts would have to expend 2,284 hours each year, an average of

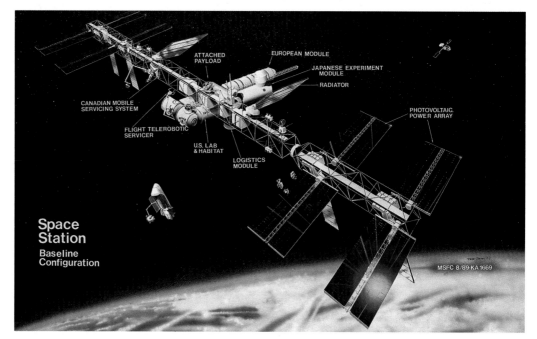

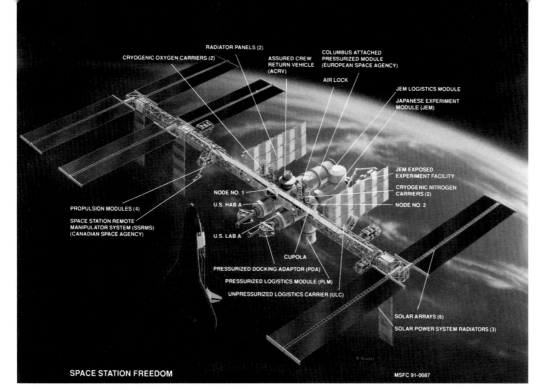

Labels on image:
RADIATOR PANELS (2)
CRYOGENIC OXYGEN CARRIERS (2)
ASSURED CREW RETURN VEHICLE (ACRV)
COLUMBUS ATTACHED PRESSURIZED MODULE (EUROPEAN SPACE AGENCY)
AIR LOCK
JEM LOGISTICS MODULE
JAPANESE EXPERIMENT MODULE (JEM)
JEM EXPOSED EXPERIMENT FACILITY
CRYOGENIC NITROGEN CARRIERS (2)
NODE NO. 2
NODE NO. 1
U.S. HAB A
PROPULSION MODULES (4)
SPACE STATION REMOTE MANIPULATOR SYSTEM (SSRMS) (CANADIAN SPACE AGENCY)
U.S. LAB A
CUPOLA
PRESSURIZED DOCKING ADAPTOR (PDA)
PRESSURIZED LOGISTICS MODULE (PLM)
UNPRESSURIZED LOGISTICS CARRIER (ULC)
SOLAR ARRAYS (6)
SOLAR POWER SYSTEM RADIATORS (3)
SPACE STATION FREEDOM
MSFC 91-0087

LEFT By 1991 Space Station Freedom had settled into the configuration concept it would eventually become. *(NASA)*

one two-person EVA every other day. By July the estimates for maintenance alone exceeded 3,276 hours per year – five two-person EVAs per week – an impossibly unacceptable target. The Johnson Space Center set about a rigid series of changes to drive annual EVA maintenance down to just 507 hours.

On 21 March 1991, NASA sent its new plan to Congress. Overall station span would be reduced from 493ft (150m) to 353ft (108m) and each of the two US modules would be downsized in length from 44ft (13.4m) to 27ft (8.2m). Whereas the modules were to have been fitted out in space, the smaller size would allow them to be assembled and launched as complete elements. They would be connected by nodes, barrel-shaped multiple docking ports serving as passageways and outfitted with equipment essential to running the station. An airlock module would also be attached to one of the node ports to allow astronauts to conduct EVA without depressurising the entire station. Instead of constructing the long truss assembly in orbit from a bundle of parts carried to orbit by the Shuttle, truss sections would now be preassembled and fitted out with conduits and cable trays on the ground, for launch as completed elements. This alone would cut EVA time by 50 per cent, requiring the crew to hook up the connections rather than erect the structure in space.

The number of Shuttle flights necessary to build Freedom had been cut from 34 to 17

but the date of the first element launch had been put back by one year to March 1996. Man-tended capability would be reached in December 1996, with Freedom complete and permanently manned with a reduced crew of four by late 1999. Already, some 79,000 workers were employed by US companies to build the various elements across 39 US states, added to which were aerospace workers in Europe, Canada, and Japan. But when Bill Clinton entered the White House in January 1993 he ordered NASA to redesign the station yet again! Over the next several months a blue-ribbon review committee set up on 10 March under former Apollo executive Dr Joseph Shea examined a wide range of configuration designs that had not been looked at before – designs that substantially downgraded the station and its capabilities to fit within budget limits.

On 17 June the president made a public statement backing a low-cost configuration known as 'Space Station Alpha' and on 24 June Clinton ordered NASA to develop a new programme plan. But on 2 September Al Gore, the vice president, sat down with the prime minister of Russia, Viktor Chernomyrdin, and signed an agreement embarking on a joint effort to design and build a completely new International Space Station. The origin of this transformation goes back to 1986, just two years after the decision by Ronald Reagan to give NASA the go-ahead to build a station.

Chapter One

A permanent place in space

While American astronauts went to the Moon and US engineers built a reusable Shuttle, Russia's cosmonauts were learning how to live and work in space. Since the early 1970s the Soviet space programme had assembled an impressive record in long-duration flight – something NASA was desperate to emulate.

OPPOSITE Mir grew over its 15-year life with additional modules and equipment, providing a base on which both US and Russian crewmembers learned how to work in space together. *(NASA)*

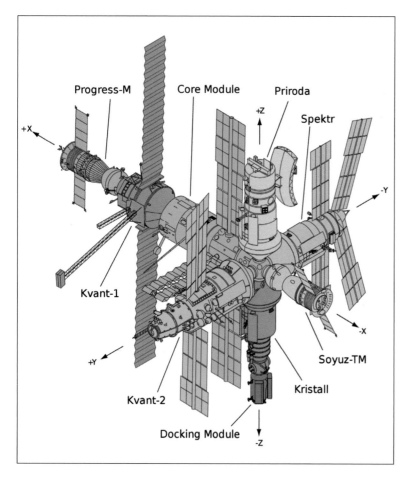

Progress-M   Core Module   Priroda

Spektr

+Z

+X

Kvant-1

-Y

-X

Soyuz-TM

Kvant-2

Kristall

Docking Module

-Z

+Y

ABOVE **Mir provided the engineering experience for Russia's contribution to the ISS and much was learned by NASA from the experiences of Russian engineers and cosmonauts.** *(NASA)*

BELOW **Salyut 7 and Cosmos 1686 provided valuable experience in docking very large structures in space.** *(NASA)*

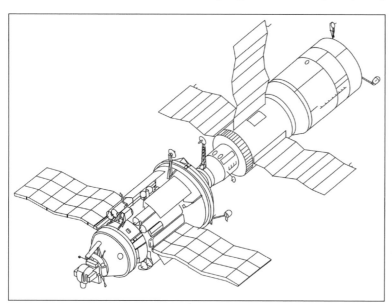

Mir

The launch of the Soviet space station Mir ('peace') came on 19 February 1986, just over three weeks after the loss of *Challenger* in a disaster that kept the Shuttle grounded until 29 September 1988. The Mir core module had been developed out of the Salyut series of space stations first launched in 1971, developed as a long-duration facility launched by a Proton rocket. Mir presaged a new capability for keeping people in space, using the building-block approach to expand the facility into a fully equipped laboratory.

To achieve that, while looking similar to the second generation stations Salyut 6 and 7 first launched in 1977, Mir had a forward docking section incorporating five ports – one at the extreme forward end and four radial ports at 90-degree intervals around the circumference. This docking section incorporated a transfer system called Lyappa, which would latch on to a new module docked at the extreme forward port and rotate it around 90 degrees, reattaching it to one of the four radial ports, thus freeing the forward port for another module.

Up to four permanent plug-in modules could be delivered to the forward node, each moved round to a radial docking ring. And just like Salyut, there was another docking port at the rear of the Mir station making six in all in a system that could accommodate four new large experiment modules while retaining respective docking ports for manned Soyuz and unmanned Progress cargo-tankers.

Internally, Mir had a single pressurised work compartment of 1,400ft^3 (40m^3), a length of 25ft 2in (7.7m), and a diameter of 9.5ft (2.9m) in the forward section and 13.75ft (4.19m) in the main body of the compartment. The rear of Mir comprised a service propulsion section with the same diameter as the aft work compartment and a length of 7.5ft (2.29m). It supported two 660lb (300kg) thrust rocket motors for re-boosting altitude and making minor orbit changes. Access into Mir from this end was made via a central tunnel 6.5ft (1.98m) in diameter connecting the aft docking port to the main work compartment.

Total length of the core module as launched was 43ft (13.1m) with a mass of 46,000lb

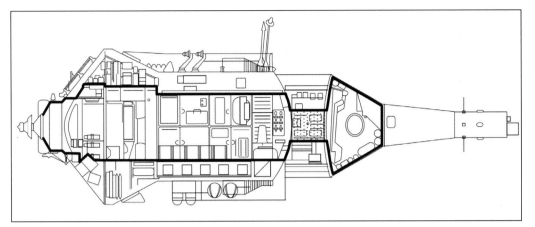

(20,865kg) and attitude control handled by 32 control thrusters. About 9kW of electrical power came from two solar array panels with a total span of 97.5ft (30m) covering 820ft^2 (76m^2), 26 per cent more than arrays on Salyut 7. Added to which, each new module would carry its own solar cell arrays to supplement power for the science equipment they contained.

The first visitors to Mir were cosmonauts Leonid Kizim and Vladimir Solovyov, launched on 13 March 1986, followed six days later by an unmanned Progress cargo-tanker which remained docked at the station until 20 April when it was de-orbited, the first of four Progress cargo-tankers launched to Mir over the next 12 months. After 50 days aboard Mir, in early May Kizim and Solovyov used their Soyuz T-15 spacecraft to fly across to the Salyut 7 space station about 1,700 miles (2,736km) away in an adjacent orbit, and became the first crew to visit two space stations on the same mission. For some 50 days more they carried out tests and retrieved some equipment before returning to Mir on 27 June. After four months in space Kizim and Solovyov returned to Earth on 16 July 1986, leaving Mir unmanned for more than six months.

Manned operations resumed on 5 February 1987, with the launch of Romanenko and Laveykin aboard Soyuz TM-2 followed by the first plug-in module, Kvant-1 (Quantum-1) on 31 March. Kvant-1 was designed as an astrophysical laboratory with various telescopes developed in cooperation with the UK, the Netherlands, West Germany and the European Space Agency. It had a length of 17.4ft (5.26m), a diameter of 14.3ft (4.34m) and weighed about 45,400lb (20,600kg) at launch. After some

difficulty getting a proper connection, Kvant-1 docked to the aft port on 9 April using an automated approach system where it would remain for the life of the Mir station, providing an additional internal volume of 620ft^3 (17.56m^3). A 20ft (6.1m) long propulsion unit, located on the extreme aft end of Kvant-1 and used only for the rendezvous manoeuvres, was separated, reducing the weight of the module to 21,000lb (9,525kg) and exposing a standard docking unit as the new aft port for future Progress and Soyuz vehicles.

For more than two years the station was manned continuously by five more Soyuz crews in succession, including cosmonauts from Syria, Afghanistan, and France, until 26 April 1989, when it was once again left unmanned. In those six expeditions, two-man cosmonaut teams spent an average of four to six months in space before returning a few days after they

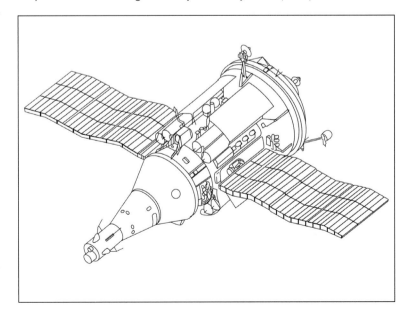

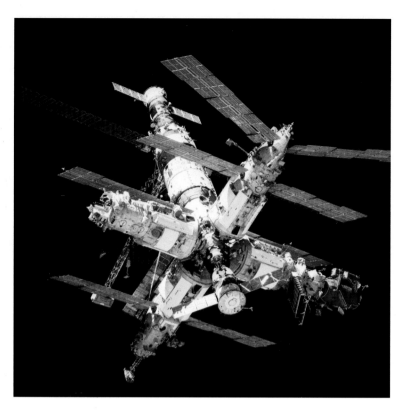

space flight were conducted at the end of 1991 and on 17 June 1992, President George Bush of the USA and President Boris Yeltsin of Russia signed an agreement pledging cooperation through Shuttle missions to Mir, the initial protocol being signed by NASA and the Russian Space Agency, the RSA, on 5 October 1992. This included the flight of a Russian cosmonaut aboard the Shuttle in 1994, a US astronaut flying on a Soyuz to spend more than 90 days aboard Mir, Russian cosmonauts being changed out via the Shuttle and development of a common (androgynous) docking module for use by either spacecraft. Throughout 1992, NASA and its industry partners began to think of how they could exploit the relaxation of political tension with Russia and utilise their considerable stock and hard-won experience in cutting the costs of the US Freedom station.

The International Space Station

were relieved by the next team. During that period too, crews offloaded equipment, fuel and water from 15 Progress vehicles that had come and gone at the aft port. Meanwhile, Salyut 7 would remain derelict, eventually burning up in the atmosphere in February 1991 and scattering debris over Argentina and Chile. A resumption of flights to the Mir station began with the flight of Aleksandr Viktorenko and Aleksandr Serebrov on 5 September 1989, little more than four months after the first crew returned to Earth, beginning a series of flights that would have the station permanently manned until 27 August 1999.

In that time, the world changed a great deal and what had begun as a race for technological and ideological supremacy between global superpowers would end with Russia free from communism. After the collapse of the Soviet empire Russia lost territory, natural resources, manpower, factories, and government facilities now located in what, almost overnight, became foreign countries. Not least was the massive former Soviet launch facility at Baikonur in Kazakhstan, future use of which was secured under a lease. There was just enough money to maintain the Mir programme but their shuttle *Buran* was abandoned.

Initial discussions to cooperate in human

The Russians had been launching stations for 20 years, routinely keeping people in space for between six and eleven months. In the end it was a presidential decision, Bill Clinton wishing to use NASA as a winch to haul in Russian expertise and forge a new working relationship. When the final agreement was signed on 2 September 1993, it probably saved the NASA station and certainly gave it new purpose. The challenge now was staggering in its implication. With a five-team partnership comprising the USA, Russia, Europe, Japan, and Canada, technical integration of separate modules and systems would be a major challenge. There were new languages, fresh working practices and very different engineering approaches, as Russia leapfrogged the junior partners and got straight down to business with the Americans. The names Freedom and Alpha One were defunct; now it was truly the International Space Station – the ISS.

On 7 December 1993, all partners formally secured the agreement for Russia to be a part of the ISS. Russian elements were added to the existing Alpha One concept and modified Soyuz would double as lifeboats, with Progress cargo-tankers supporting what was now an enlarged operational plan for sustained use. Perhaps the biggest change of all was to place the station in an orbit with a greater inclination to the equator – 51.6 degrees instead of 28.5 degrees – and there were two reasons for this. Russia would launch from the Baikonur complex in Kazakhstan at latitude 45.6 degrees N. Sending a launch vehicle to a lower inclination would cost it in payload capability. Also, it had long been argued by Canada, Europe, and Japan that their geographic locations were far to the north of the earlier inclination of 28.5 degrees and that Earth observations would be compromised. Under the new plan assembly of the ISS would begin in November 1997, prior to which NASA planned to fly up to ten Shuttle missions to Mir carrying cargo, an airlock module, and supplies.

By the time the agreement with Russia was signed, Mir had begun to grow. Kvant-2 was launched on 26 November 1989, dedicated to Earth observations but also carrying a shower, oxygen production electrolysis equipment, food and propellant supplies, and a wide range of supplementary items. With a length of 45ft (13.7m) and a diameter of 14.3ft (4.34m), Kvant-2 had a mass of 40,800lb (18,500kg) and was the first plug-in module manipulated from the forward port to which it had initially docked to one offset 90 degrees to the side using

the Lyappa system. It had two solar arrays providing 6.9kW of electrical power and in its three sections – airlock, central science area, and aft equipment section – had an internal pressurised volume of 930ft^3 (26.33m^3), 50 per cent more than Kvant-1.

ABOVE The configuration of the ISS was fixed when the Russians became a part of the team, the central truss supporting modules provided by Russia, the USA, Europe and Japan, Canadian telerobotic systems being added as the station grew. *(NASA)*

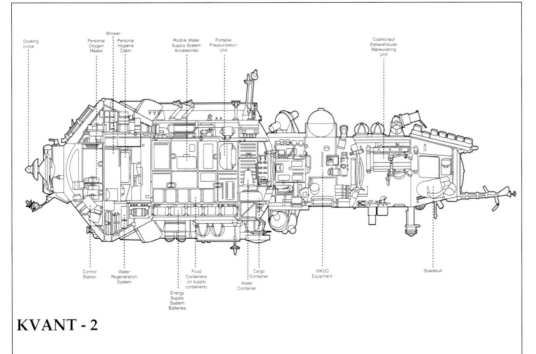

KVANT - 2

LEFT Launched in 1989, Kvant-2 was the first module based on the TKS design and carried a high resolution camera, spectrometers for earth observation and astrophysical experiments. *(NASA)*

ABOVE The two rear modules were built and launched by Russia, and form the base on which Soyuz and Progress vehicles are docked. *(NASA)*

In less than three years the station had grown from 1,400ft³ (39.64m³) to 2,950ft³ (83.53m³). This was raised to 3,895ft³ (110.29m³) with the launch of the Kristall materials processing and biotechnology laboratory on 31 May 1990. Docked initially to the front port but moved to a radial port, Kristall had the same dimensions as Kvant-2, with a weight of 43,300lb (19,640kg), of which 15,400lb (6,985kg) was cargo for Mir. The pressurised internal volume of 945ft³ (26.76m³) was divided into two sections by a circular 31in (79cm) hatch. Some 5.5–8.4kW of electrical power came from two concertina-like solar arrays and had great flexibility. At one

point a wing would be removed and relocated to less shaded positions on Kvant-1, the first relocatable solar array, and the Kristall module itself would be moved around docking ports no less than four times during 1995 for operational reasons.

Between 1993 and 1995 the new plan for joint operations took form. Russia would receive more than $1 billion for its contribution to the ISS programme and NASA would fly supplies to the Russian space station. Spektr and Priroda, two new modules launched by Russia to Mir in 1995 and 1996 respectively, would carry US instruments for materials processing, biotechnology, Earth sciences, and space technology experiments. The flight of an American aboard Mir in 1994 and the flights by Shuttle became known as the Phase I programme, with assembly of the ISS itself becoming Phase II. There would be two Mission Control Centres: MCC-H in Houston in charge of the Shuttle and MMC-M in Moscow in command of Mir. Neither one would have overall command of the mission. Joint flight rules were, however, developed for use in specific situations and special consoles were set up in each facility to interact in decisions between these two facilities, which were almost 6,000 miles (9,650km) apart.

Because it was not possible within the

RIGHT Added to Mir in 1990, the Kristall materials processing laboratory conducted research on materials, the production of new semiconductor crystals and research on astronomy and astrophysics. *(NASA)*

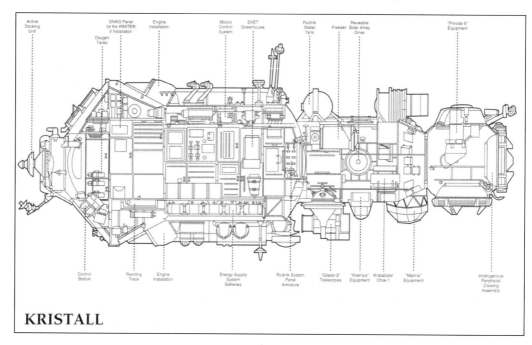

KRISTALL

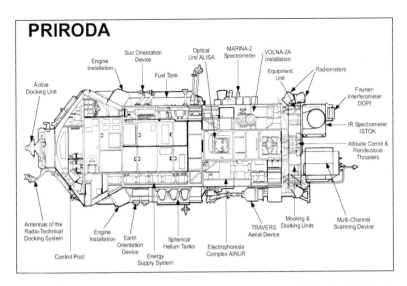

PRIRODA

Sun Orientation Device · Engine Installation · Fuel Tank · Optical Unit ALISA · MARINA-2 Spectrometer · VOLNA-2A Installation · Equipment Unit · Radiometers · Active Docking Unit · Founer-Interferometer DOPI · IR Spectrometer ISTOK · Attitude Control & Rendezvous Thrusters · Antennas of the Radio-Technical Docking System · Engine Installation · Earth Orientation Device · Energy Supply System · Spherical Helium Tanks · Electrophoresis Complex AINUR · TRAVERS Aerial Device · Mooring & Docking Units · Multi-Channel Scanning Device · Control Post

RIGHT With dedicated remote sensing and earth observation equipment, Priroda was added to Mir in 1996. *(NASA)*

CENTRE Added to Mir in 1995, Spektr conducted remote sensing of the earth with solar arrays that virtually doubled the electrical production capacity of the station. *(NASA)*

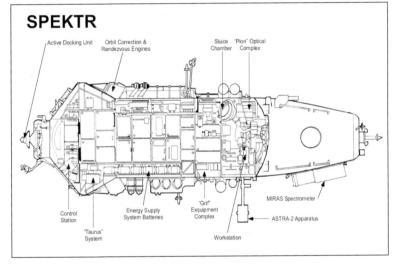

SPEKTR

Active Docking Unit · Orbit Correction & Rendezvous Engines · Sluice Chamber · "Pion" Optical Complex · Control Station · "Taurus" System · Energy Supply System Batteries · "Grif" Exquipment Complex · Workstation · MIRAS Spectrometer · ASTRA-2 Apparatus

limitations of Phase I for every crew member of one nationality to speak fluently the language of the other, dual-language checklists and documents were used. Another problem was sleep. Long-duration Mir crews worked a normal 24-hour day aligned with Moscow time but short-duration Shuttle crews slipped their 'days' according to launch and landing times. For those flights where astronauts or cosmonauts would be carried as guests in respective spacecraft, training in each other's facilities was essential. Given that crewmembers were already seasoned astronauts, each was required to go through 800–900 hours of training to prepare for their flights, including practising for EVAs in suits, for which some degree of familiarity was essential. The Russian Orlan suit was very different to the US Extravehicular Mobility Unit (EMU) and full rehearsal in a neutral-buoyancy water tank was essential. In all, seven joint Shuttle–Mir EVAs would be carried out.

Another training and orientation requirement was to familiarise themselves with each other's spacecraft. Soyuz was very different to the Shuttle and flew a ballistic rather than a lifting re-entry, imposing much greater g-forces on its occupants during both launch and descent than did the benign Shuttle. In the Shuttle forces rarely reached 3g but Soyuz would routinely impose up to 9g and while this is not excessive for trained personnel, the discomfort of a cramped cabin and a somewhat more severe re-entry could be challenging, especially after long periods of weightlessness.

RIGHT Originating in a design put together in 1963, the Soyuz spacecraft has proved to be a reliable means of transport to and from Salyut stations, Mir and now the ISS. *(NASA)*

Chapter Two

ISS Phase One – missions to Mir

Following agreement to merge their efforts at long-duration missions in Earth orbit, NASA and its Russian counterpart set to work planning a series of missions that would prepare the way for assembly of a truly international laboratory in space – one hosting astronauts from several different countries.

OPPOSITE An artist depicts the docking of *Atlantis* to the incomplete Mir in November 1995. *(NASA)*

The plan to merge the ambitious human space flight programmes of the United States and Russia was both brave and bold. Brave, because it faced some objections from European countries which, until the decision in 1993 to embrace the newly democratised Russia, had been America's prime partner through its European Space Agency; now Russia was the main partner and the others, including Japan, felt marginalised. And it was bold because the way the Russians went about engineering their space vehicles was completely different to that of the United States; there would have to be much compromise and a lot of information exchange from each side to the other.

In truth each had much to learn from the other and that had been understood for the previous 20 years, when the Apollo-Soyuz Test Project (ASTP) brought astronauts and cosmonauts together for a joint docking flight in a mission launched on 15 July 1975. Just over two days later they docked, visited each others' spacecraft and spent almost two days carrying out joint operations. This flight, accomplished during the Cold War, inspired plans for a mission to link NASA's Shuttle with a Russian Salyut station, but politics got in the way and it never happened.

Now, Phase I of the 1993 agreement would involve Shuttle flights to the latest Russian station, Mir. In that respect it was the next logical step beyond ASTP and the realisation of a plan that advocates on both sides of the political divide had wanted since that first joint endeavour. ASTP had given NASA the opportunity to see for itself how the Russians worked, solved problems and constructed solutions, and each side gained a new and healthy respect for the other's work. Now it was time to put all that to the test. But only by flying together in space could the real test take place.

STS-60

3–11 February 1994

The historic flight of NASA's sixtieth Shuttle mission was a landmark event, being the first US spacecraft to carry a Russian cosmonaut as part of its crew. But it would not rendezvous or dock with Mir. Launched on February 3 1994, STS-60 was commanded by Charles F. Bolden and included in its six-person crew Sergei Krikalev, who participated in this purely scientific flight. During the eight-day mission experiments were conducted, and a commercially developed Spacehab module attached to the forward part of the payload bay allowed several other tasks to be conducted with a special video-link set up between the Shuttle and three cosmonauts aboard Mir.

STS-63

3–11 February 1995

Launched exactly a year after the first US flight carrying a Russian into space, this was the first flight to carry a female pilot – Eileen Collins – and to include the STS-60 back-up cosmonaut Vladimir Titov. The eight-day flight included the first rendezvous with the Mir space station, *Discovery* coming to within 36ft (11m) of the orbiting complex before backing away to a distance of 400ft (122m) to commence a slow fly-around for a photo-survey. A Spacehab module once again provided opportunity for science and technology tests and a set of six tiny sub-satellites with which ground controllers could calibrate radar equipment to more precisely monitor space debris were released. Other tests into how the Shuttle environment affects the space around it were conducted

BELOW Sergei Krikalev (top right) became the first Russian to ride the Shuttle when STS-60 was launched in February 1994. Charles F. Bolden (bottom right) became NASA Administrator in 2009. *(NASA)*

using a SPARTAN platform deployed by the remote manipulator. British-born Michael Foale joined Bernard Harris on a 4hr 38min EVA, Harris becoming the first Afro-American to walk in space.

MIR-18/NASA-1

14 March 1995

Little over a month after *Discovery* returned to Earth, the first US astronaut to ride a Russian spacecraft was launched from Baikonur when Norman Thagard, the first American to visit the Russian station, was carried to Mir along with cosmonauts Vladimir Dezhurov and Gennady Strekalov aboard Soyuz TM-21. They joined Aleksandr Viktorenko and Yelena Kondakova, who had been aboard Mir since October 1994 but would return in Soyuz on 22 March. The unmanned Spektr module launched by Russian rocket on 20 May 1995 carried US instruments and docked to the forward Mir port on 1 June. With a weight of 43,300lb (19,640kg), Spektr was 47.4ft (14.5m) long and had a diameter of 13.45ft (4.1m), carrying remote sensing instruments for Earth and atmospheric studies. Spektr would also provide living and working areas for visiting NASA astronauts. It had two Kvant-2 type solar arrays with a total span of 76.45ft (23.3m) and produced 6.9kW of electrical power. Spektr was adapted

from an abandoned military programme for space surveillance and missile defence and the forward Octava module was replaced with a conical device. Both Spektr and the later Priroda modules were paid for by NASA and equipped with US instruments. Before Spektr docked to Mir the two Russian cosmonauts conducted three spacewalks to complete installation of Kristall's solar wing on Kvant-1 and retract Kristall's second solar wing to make way for *Atlantis* arriving a month later.

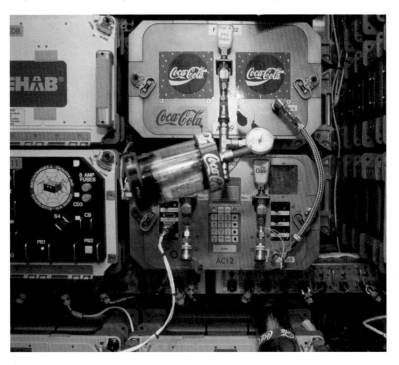

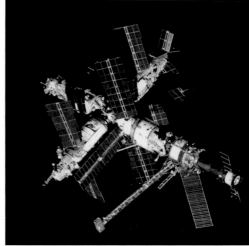

STS-71

27 June–7 July 1995

The first Shuttle to dock with Mir delivered cosmonauts Anatoly Solovyev and Nikolai Budarin and returned ten days later with Thagard, Dezhurov and Strekalov. Thus it fell to *Atlantis* to be the first NASA spacecraft to return with a partially different crew to that which it had carried on launch day and for this mission the resident Shuttle crew going both ways were NASA astronauts Robert 'Hoot' Gibson, Charles Precourt, Ellen Baker, and Bonnie Dunbar. On the approach to Mir, Gibson was at the controls and conducted the approach from below the station, holding formation with Mir at a distance of 250ft (76m) for the vital 'go' from mission control centres in Houston and Moscow. Tension was high as Gibson eased the two together at a speed of 0.7mph (1.13kph), 250 miles (400km) above Lake Baikal during the afternoon local time on 29 June. The docking unit was an androgynous system developed

specifically to connect the airlock module in the forward Shuttle cargo bay with the docking port on the Kristall module. Using the Lyappa system, on 10 June Kristall had been moved from its radial position to the forward port on Mir so that *Atlantis* could dock along the long axis of the station rather than off to one side, giving it clearance over solar arrays.

To bring the experience closer to Earthlings, *Atlantis* carried an IMAX camera for cinema-style filming, and five schools in the US had dedicated links to the docked complex for question and answer sessions that would soon become popular on Shuttle flights. Some 15 experiments in human biology and life sciences kept the ten crewmembers busy until undocking on 4 June, prior to which Solovyov and Budarin climbed into their Soyuz spacecraft and separated from Mir to photograph the historic departure before returning to begin a long stay in space. Returning with a record eight crewmembers, *Atlantis* had cleared the way for routine Shuttle–Mir operations and brought astronaut Thagard back from his 115 days in space, extending in duration the 84.5-day US flight record of the third and final Skylab crew, set more than 21 years before.

STS-74

12–20 November 1995

For the second Shuttle–Mir docking mission *Atlantis* carried a new 9,000lb (4,080kg) docking module (DM), 15.4ft (4.7m) tall and 7.2ft (2.2m) in diameter. Development of the DM had been fast-tracked by the need to

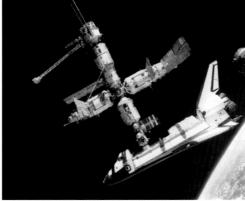

have a module that could extend the distance between the Shuttle and the Kristall module so as not to interfere with solar cells on the latter, Kristall having been rotated back to its radial docking position on 17 July. To achieve a link-up, the DM had first to be lifted from the rear of the Shuttle's cargo bay to the top of the forward airlock at the front of the bay. In addition to two strap-on solar cell arrays that would be positioned on Mir during a subsequent spacewalk, *Atlantis* also provided 1,000lb (454kg) of water for the Russian station. Shuttle crewmember Chris Hadfield was a Canadian and on 3 September the Soyuz TM-22 spacecraft had delivered cosmonauts Yuri Gidzenko and Sergei Avdeyev and German astronaut Thomas Reiter from the European Space Agency to the station awaiting the arrival of STS-74. Thus Mir hosted crewmembers from all the participating ISS countries with the exception of Japan. When *Atlantis* and its five-man crew separated from Mir on 18 November it left the DM attached to the Kristall module for later Shuttle missions.

STS-76

22–31 March 1996

The third Shuttle–Mir docking delivered Shannon Lucid, the first female NASA astronaut aboard the station, at the start of a planned 142 days in space that would see the first EVA conducted during a Shuttle–Mir docking. The two NASA astronauts assigned to the EVA, Linda Clifford and Rich Godwin, exited via the Shuttle airlock hatch on 27 March to install sets of experiments on the outside of the Mir station and test a variety of tethers, restraints, and special tools that would be used with the ISS. Undocking came on 24 March, Shannon Lucid remaining with cosmonauts Yuri Onufrienko and Yuri Usachev, who were previously launched to Mir aboard Soyuz TM-23 on 21 February. As a historical footnote, this was NASA's last human mission managed from the original Mission Control Center at the Johnson Space Center, Houston. After five days docked to Mir, *Atlantis* separated on 29 March and returned to Earth two days later.

The last of five modules flown to Mir was launched on 23 April 1996, when a Proton

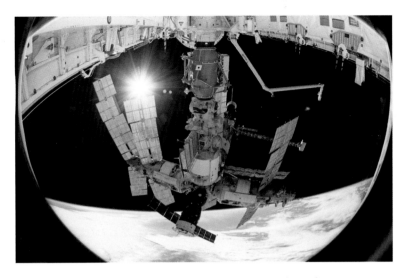

ABOVE From the aft end of the Shuttle cargo bay, a fish eye lens captures *Atlantis* docked to the Mir station on STS-74. *(NASA)*

rocket lifted the 43,400lb (19,685kg) Priroda laboratory into space. It docked to the forward axial port three days later and was transferred to the last remaining side position by the Lyappa system on 27 April. With a length of 42.6ft (13m) and a diameter of 14.3ft (4.36m), Priroda was packed full of equipment not only from Russia but also contributed by Bulgaria,

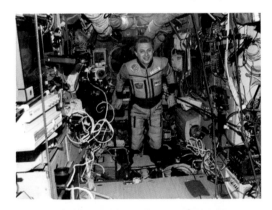

LEFT Cosmonaut Yuri Onufrienko shows off the interior of the Mir core module. *(NASA)*

BELOW STS-76 delivered Shannon Lucid, the first female cosmonaut to fly aboard Mir. *(NASA)*

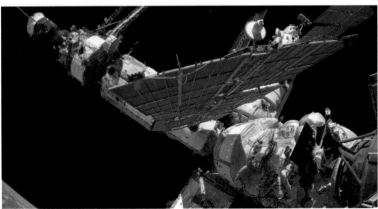

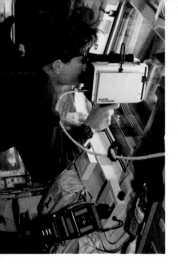

ABOVE Linda Godwin points a laser range finder at the Mir station during rendezvous operations on STS-76. *(NASA)*

ABOVE RIGHT Kvant-1 supports Soyuz TM-23 which had been launched February 21, 1996, and remained with Mir until September 2. *(NASA)*

Germany, Poland, Romania, the Czech Republic, and the USA covering materials science, biotechnology, life sciences, Earth observation, and space technology.

The next Shuttle mission should have launched on 31 July 1996, replacing Shannon Lucid with John Blaha, but that mission was delayed when booster problems kept *Atlantis* grounded. On 17 August Soyuz TM-24 carried cosmonauts Valery Korzun and Aleksandr Kaleri and French astronaut Claudie Andre-Deshays to Mir for a docking, now placing two women aboard Mir for the first time. She returned to Earth on September 2 with Onufrienko and Usachev in the Soyuz T-23 spacecraft, leaving Shannon Lucid with her two new Russian colleagues until *Atlantis* could be cleared to collect her.

STS-79

16–26 September 1996

Heading for the fourth Shuttle–Mir docking, despite delays *Atlantis* finally got off the pad for a ten-day mission with a six-man crew commanded by William Readdy. Docking occurred on 19 September and *Atlantis* undocked four days later having collected Shannon Lucid, depositing John Blaha for his long-duration flight aboard Mir. In space for 188 days – 46 days longer than planned due to Shuttle delays – Lucid, a veteran of four previous Shuttle missions, now had the double record of longest space flight for both an American and a woman. Her record stood for more than ten years until it was broken by Sunita Williams on 16 June 2007, aboard the ISS. *Atlantis* delivered a record 4,600lb (2,087kg) of water, food, and other items to Mir and brought back 2,200lb (1,000kg) of scientific equipment. This was the first flight to carry the 15,500lb (7,000kg) Spacehab double module in the Shuttle's cargo bay, which served as a store for hardware up to Mir and down to Earth. Because Blaha was in orbit during the 1996 US elections he was prevented from casting his vote, a dilemma that a year later resulted in the Texas legislature changing its statutes to permit voting from space!

RIGHT NASA astronaut Shannon Lucid floats in the Mir space station. She was returned to Earth by STS-79 having spent 188 days 4hr 14s in space. *(NASA)*

BELOW STS-79 was the fourth Shuttle mission to dock with Mir, placing John Blaha on board for the third of NASA's long duration stays aboard the station. *(NASA)*

RIGHT Shannon Lucid moves her space suit back into the Shuttle prior to returning home. *(NASA)*

STS-81

12–22 January 1997

Commanded by Michael Baker, the fifth Shuttle flight to Mir offloaded 6,000lb (2,720kg) from *Atlantis* and downloaded 2,400lb (1,090kg) of equipment, returning John Blaha to Earth after his 118-day flight following its five days docked to the Russian station. *Atlantis* delivered Jerry Linenger to Mir at the start of his stay, the third US long-duration mission aboard the station. As on previous visits, valuable working practices, methods, and procedures that would be essential for integrating international crewmembers aboard the upcoming ISS were explored as well as solutions found to unexpected operating difficulties.

Almost three weeks after *Atlantis* returned to Earth, ESA astronaut Reinhold Ewald accompanied cosmonauts Vasili Tsibilev and Aleksandr Lazutkin aboard Soyuz TM-25, launched to Mir on 10 February. The automatic docking on 12 February failed and a manual approach was necessary, after which the three crewmembers joined Korzun, Kaleri and Linenger aboard Mir. But on 23 February a fire broke out when a lithium hydroxide canister exploded. Quickly extinguished, the fire produced toxic gases and the crew wore breathing gear until the fumes dispersed through the air recirculation system.

Then, two days after Korzun, Kaleri, and Ewald returned to Earth in the Soyuz TM-24 spacecraft on 2 March, Progress tanker M-33 was unable to dock and the attempt was abandoned. The Progress spacecraft was a one-way delivery truck and so M-33 had to be sent to a fiery end in the atmosphere on 12 March, packed full of supplies it had been unable to deliver. Finally, on 8 April, Progress M-34 successfully docked at the aft port on Kvant-1 and its supplies offloaded, including a new Elektron oxygen production unit.

STS-84

15–24 May 1997

Mindful of the fire and the failed Progress link-up, it was with some trepidation that *Atlantis* was launched for the sixth Mir docking carrying a crew of seven. British-born astronaut Michael Foale replaced Jerry Linenger for NASA's next long-duration stay aboard the

station. Linenger had logged 132 days in orbit, at that time the longest of any US astronaut, and during his tenure had become the first American to conduct an EVA from a non-US spacecraft wearing a Russian Orlan space suit. Linenger had experienced a near-catastrophic fire, a failure in the carbon dioxide filter system, a failure in the Mir attitude control system and the near collision by a Progress vehicle that failed to dock with the station. And then it got worse.

On 24 June, a month after *Atlantis* returned to Earth, Progress M-34 was undocked for relocation and while manoeuvring around the station it went out of control, crashing into the Spektr module, causing depressurisation and smashing solar arrays. Scurrying to seal off the module, the crew isolated Spektr and saved the rest of the station complex. Three days after the collision batteries on Kvant-2 ran down and TM-25's thrusters were used to re-boost the station's altitude. Progress M-35 arrived on 7 July and docked to the Kvant-2 port, bringing fresh supplies. A week later cosmonaut Tsibilev showed signs of cardiac arrhythmia and his workload had to be drastically reduced.

On 5 August, Soyuz TM-26 carried cosmonauts Anatoly Solovyov and Pavel Vinogradov to Mir where they became Foale's new companions, with the resident crew of Tsibilev and Lazutkin returning to Earth in their TM-25 spacecraft on 14 August. The next day the Mir crew flew around the station in TM-26 to view the damage and assess the condition of the complex. This was followed by a spacewalk conducted by Solovyov and Vinogradov on 22 August to cut Spektr's severed electrical cables and rewire remaining solar panels, restoring 70 per cent of the original capacity. But Spektr itself remained sealed. And then Mir suffered a major computer failure, which was unresolved by the time the crew got it started again. The launch of Progress M-36 was consequently delayed so that a replacement could be carried aloft.

STS-86

25 September–6 October 1997

The seventh Shuttle mission to Mir was the last visit for *Atlantis*, which on this flight delivered 7,000lb (3,175kg) of supplies including 1,700lb (770kg) of water. For five

days the crew conducted a wide range of experiments, returning with Michael Foale, who had spent 144 days in space, and leaving David Wolff in his place with cosmonauts Solovyov and Vinogradov. Diminutive astronaut Wendy Lawrence was supposed to have replaced Foale aboard Mir but concerns about the minimum size Orlan EVA suit requirement prevented her from flying this mission. For the remainder of the year Mir struggled on with several failures to computers and attitude control systems. Two Progress cargo-tankers delivered supplies and stores as required.

STS-89

22–31 January 1998

The penultimate Shuttle–Mir docking saw *Endeavour* – the replacement for *Challenger*, which was destroyed on 31 January 1986 – make its 12th flight, carrying a crew of seven plus 7,000lb (3,175kg) of stores for the station, docking on 24 January. Astronaut Andrew Thomas replaced David Wolff, who had logged 128 days on his long-duration mission aboard Mir. Undocking came after almost five days of dual activity, Wolff remaining aboard Mir with Solovyov and Vinogradov on what was the last US long-duration visit to the station before translating to the International Space Station, the assembly of which would begin later the same year. For the next four months Mir's resident crew of Talgat Musabeyev and Nikolai Budarin would maintain the station and carry out research through the last Shuttle visit in June. Delivered to Mir after launch by Soyuz TM-27 on 29 January, they came up

with French astronaut Leopold Eyharts, who returned with Solovyov and Vinogradov in Soyuz TM-26 on 19 February.

STS-91

2–12 June 1998

Andrew Thomas had completed 141 days in space when he joined the six-person crew aboard *Discovery* and returned to Earth at the end of the last Shuttle–Mir mission. Commanded by Charles Precourt, STS-91 included in its crew Wendy Lawrence, unable to fly STS-86 due to suit undersize restrictions. The Shuttle had arrived with about 5,800lb (2,630kg) of water and stores and carried a prototype Alpha Magnetic Spectrometer designed to look for dark matter in the universe. Docked for almost four days, *Discovery* departed on 8 June, the station now operated in its last two years solely by Russian cosmonauts in space and controllers on the ground.

Mir made redundant

Russia's Mir space station had been the pride of the Soviet Union but when the communist regime collapsed it was a financial burden the country was no longer able to afford. Had not the agreement been signed with the United States in 1993, Russia would have been unable to continue with Mir and probably would by now be out of the human space flight business. As a result of the agreement, in addition to launch and operational costs, the US paid $472 million for trips to the Russian space station and kept the programme alive. At least, that was, until agreement was finalised on the International Space Station, by which time the economic situation in Russia had improved. In the nine docking missions to Mir the Shuttle delivered 50,500lb (22,900kg) of cargo and equipment and returned to Earth carrying 17,200lb (7,800kg) of redundant equipment and waste, a staggering logistical transfer of 67,700lb (30,700kg). But even at the end, as the Russians kept it alive for two more years, Mir was costing that country $230 million a year just to keep it running. Nevertheless, Mir was a proud icon of a once dominant space programme and the Russians were determined to keep it operational even as they were helping construct the ISS.

After the extensive involvement of NASA and its nine Shuttle docking flights, the Russians struggled to keep Mir operational. The condition of the space station, now more than 12 years old, was declining, and sustainability of what had, in reality, been the world's first international space station was increasingly untenable. Musabeyev and Budarin continued to operate the station until they were replaced by Padalka and Avdeyev in Soyuz TM-28 on 13 August 1998. Many things were changing in Russia and the government divested its interest to a commercial operator known as MirCorp, making it the world's first commercial space platform. Paid for by an agreement between MirCorp and manufacturer Energia, a crew change in February 1999 carried Russia's Viktor Afanaseyev and France's Jean-Pierre Haignere to join Avdeyev while Padalka returned to Earth. When these three crewmembers returned home on 27 August 1999, Mir was left unmanned for the first time in almost ten years. Mir had supported human life for 4,592 days and sustained an unbroken permanent presence for 3,640 days, 22 hours, and 52 minutes since 5 September 1989.

With 60 per cent owned by Energia and 40 per cent by private investors, MirCorp sought to market Mir for space tourists. An early subscriber, American millionaire Dennis Tito, was booked to fly but the Americans resisted giving their approval. The Russians had signed an agreement with the international station partners that they would not mix valuable resources by operating Mir while developing ISS elements. Since STS-91 three Progress tankers had restocked the station but with assembly of the ISS by now well under way, the decision was made to deliberately de-orbit the venerable station. To do that a special Progress vehicle would dock to the Kvant port and, through a series of three rocket burns, bring it down over a remote region of the Pacific Ocean.

Before that, in February 2000 Progress M1-1 docked to Mir and on 6 April a new and improved Soyuz, TM-30, delivered cosmonauts Sergei Zalyotin and Aleksandr Kaleri for a final visit on what was dubbed the 'MirCorp mission'. They conducted a spacewalk, another commercial 'first', on 12 May to examine Mir's exterior and inspect an errant Solar Array Wing on the Kvant-1 module, discovering burnt wires that had cut power. The first 'commercial' Progress cargo-tanker docked with Mir in April and on 15 June Zalyotin and Kaleri returned to Earth. Having a $70 million order backlog for tourist flights, the Russians eventually succumbed to pressure from the Americans – for the time being. Just a few months later the Russians would put Tito aboard the ISS and begin a successful programme of carrying tourists to the station!

The final Progress sent to Mir, M1-5, was launched on 24 January 2001 and docked three days later. On 23 March it performed three de-orbit burns in succession, bringing Mir down to fiery destruction, seen from Fiji at 5.50pm local time. Mir had been in space for 5,519 days and made 86,331 orbits of the Earth at an average altitude of more than 200 miles (322km), travelling a total distance around Earth of more than 2 billion miles (3.22 billion km). It had been an emotional end to a decade of outstanding success. For the first time, a station launched for basic research had grown to many times its original size through the addition of specialised modules and complex equipment lifted from Earth to a permanent home in space. Along the way the station had proven the concept of automated supply and refuelling operations using Progress tankers, derivatives of the manned Soyuz spacecraft first flown in the mid-1960s and now the standard workhorse for taking people to and from low Earth orbit.

In time Soyuz would twice save the future International Space Station from being left untended when first the Shuttle stood down after a catastrophic disaster in 2003 and again after the last Shuttle came home in 2011. But Mir had itself given life to NASA's aspirations for a permanently manned facility of its own, shared with international partners encompassing long-term friends and former adversaries. It is not too extreme a view that had Mir not happened, the ISS would never have been built. Saved politically by a US president, the ISS is the very embodiment of Cold War attitudes applied to post-Cold War logic, for a cohesive and cooperative bond between technologically superior states.

Chapter Three

ISS Phase Two – assembly

What began as an idea to work together in space, assembling a giant orbiting laboratory, developed only gradually as technical and political differences intervened to slow the process. The big challenge was learning how to trust each other, former adversaries working as one for the first time.

OPPOSITE Sunrise over Zvezda, seen 16 times in one Earth day. *(NASA)*

RIGHT The various
elements of the
ISS colour-coded
according to the
country of origin,
looking more like a
child's construction
kit. (NASA)

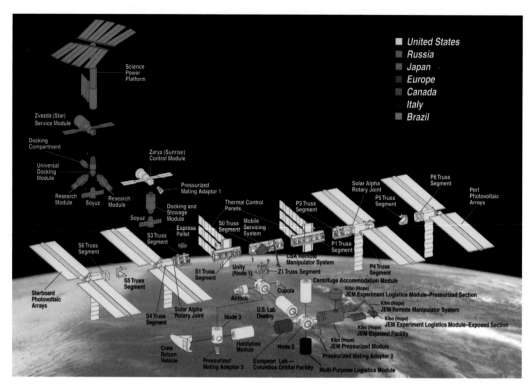

By 1995 the scale of the ISS had grown to 75 assembly flights in a complex schedule of launches planned to start in November 1997 and finish in June 2002. By 1997 the number of prime flights had been cut back to 45 and the final configuration settled. The partners finally agreed on a 'T' configuration for the ISS modules, with a single truss assembly supporting eight individually geared solar array panels, four at each end. The long axis of the 'T' modules would have a length of 167ft (51m) and the top bar of the 'T' would represent the opposing reach of the European Columbus and Japanese JEM modules located either side of the Node 2 (Harmony) module. The truss supporting the four Solar Array Wings at each end would be attached to the top of the US module and have a span of 357ft (109m). It would take more than a decade to build the ISS and because of that it would evolve incrementally, with several elements moved around to balance the geometry of the configuration but also make use of each element in a productive way. Because of this, the location of individual cargo elements would change as the station grew, complicating a straightforward explanation of how it was assembled.

A key part of the design was the need for the structure to be stable at every step in the assembly process. Although weightless, any structure in orbit has mass and inertia and the Earth's gravity tends to pull it around to point downwards like a pendulum. Also, the layout of the structure must be arranged around its centre of mass so that aligning it correctly for operational use will not pose attitude control problems. It was these unseen forces playing on all structures in space that bedevilled early configurations and experience with controlling massive objects such as NASA's Skylab and Russia's Mir were useful in tweaking the design and assembly sequence for the ISS. Another consideration arose from the length of time needed to completely assemble all the large and complex truss beams for supporting the huge solar arrays. Some of those arrays would be needed early on in the assembly phase to provide electrical power for the station and would have to be moved around to convenient locations until the final main truss structure had been assembled.

Under the new plan, the Shuttle–Mir flights of Phase 1 would be followed by Phase 2, beginning with the launch of the first element and ending with sufficient assembly to begin permanent habitation. By 1998 it was hoped to complete Phase 2 by early 2000. Phase 3 would mark completion of full assembly with all modules and structures installed, anticipated for 2004. The first element launched was the Russian-built Zarya which would be followed by a Shuttle carrying the first of three nodes called Unity, docked to one end, followed by another Russian-built module called Zvezda docked to Zarya. Together they would comprise the Russian section of the ISS. Docking ports on the Russian section would allow other modules to be attached after assembly had been completed.

Zarya/ISS 1AR

20 November 1998, 6.40am GMT

Zarya's launch was delayed beyond November 1997 when construction fell behind schedule due to financial difficulties on the Russian side. Under the original plan put together in 1994, Boeing would build the first element of the station known as Bus-1 for a cost to the US taxpayer of some $450 million. As the price of Bus-1 began to rise still higher, Boeing, under pressure from NASA, was persuaded to cancel it and pay a Russian manufacturer, the Khrunichev State Research & Production Space Centre (KhSc), to deliver a replacement named Zarya ('sunrise' in Russian), which had been the name Russian engineers had wanted to give the Salyut 1 space station almost three decades earlier.

Zarya was based on the so-called Functional Cargo Block (FGB), and was bought by the Americans for $220 million, the Russians having written off its development costs in those earlier programmes. Designed essentially for living and working in, the FGB evolved from the TKS resupply ship for the Russian Almaz military space station programme. That had been first launched as Cosmos 929 on 17 July 1997. That design formed the basis on which Mir modules were developed and was now to become the first ISS element launched into space.

Assembly of Zarya began in December 1994 and it was delivered to the Baikonur cosmodrome in January 1998 as the political wrangling over delays reached fever pitch. With problems emerging at this early stage many US politicians feared the worst, citing Russian corruption and obfuscation as reasons for America to pull out of the programme. Political pressure was applied and the funds that should have been going to produce Zvezda were

found. Zarya was launched by Proton rocket on 20 November 1998, the first element in an assembly sequence that would last more than 12 years.

Anatomy
Zarya module

Length	41.2ft (12.56m)
Diameter	13.5ft (4.11m)
Pressurised volume	2,525ft^3 (71.5m^3)
Solar array span	80ft (24.39m)
Weight	42,600lb (19,323kg) fuelled and 17,600lb (7,983kg) unfuelled
Launched	20 November 1998
Launch vehicle	Proton
Launch site	Baikonur

Visually similar to the Mir core module, Zarya had an internal pressurised compartment and a rear equipment and propulsion section and would serve initially as a powerhouse and tug. It was equipped with a Kurs automatic rendezvous and docking system and the forward section comprised a spherical compartment with single forward (axial) and radial docking ports to which Soyuz and

RIGHT STS-88
launched on
December 15, 1998,
and approached Zarya
for a docking 60 hours
47 minutes later.
(NASA)

Progress vehicles could be attached. Zarya was equipped with 24 docking and stabilisation motors, each with 88lb (40kg) thrust, and 12 small, 2.9lb (1.3kg) thrust stabilisation motors. Two rocket motors for orbital changes essential to rendezvous and docking each had a thrust of 919lb (417kg). The 12,698lb (5,760kg) of N2O4/UDMH propellant was carried in external tanks, with eight short tanks each holding 105.6 gallons (480 litres) and eight long tanks each with 87.17 gallons (396 litres). Up to 3.3kW of electrical power was produced by two solar arrays, each 35ft (10.7m) long and 11ft (3.35m) wide, and the module would provide orientation and propulsion for the early configuration of ISS during the initial stages of assembly.

In cooperation with Boeing, Honeywell installed multiplexer/demultiplexer units enhanced with computer software and Intel 80386 processing cards. The pressurised compartment contained equipment for life support, command and data handling, electric power, and docking. For attitude control several star, sun and horizon sensors provided data integrated with measurements from magnetometers, rate gyroscopes and Russia's Glonass navigation satellites. Commands from Russia's mission control could be sent through the Regul system or from Luch satellites. Inside the module, analogue communications provided voice, paging, and caution and warning signals with information received and transmitted via two audio communication units operating at VHF frequencies.

To provide electrical power the two solar arrays unfold in orbit. Each array supports 14 solar cell panels and comprises six electronically independent generators, each with 85 solar cells connected in series. Output varies from 24–34 volts. For two-thirds of each orbit Zarya would be in the Earth's shadow and the arrays would be motionless. Power to the module at these times would be provided by a battery that stores electrical energy when the arrays are in sunlight. To charge the battery and provide direct power the arrays track the sun as it arcs through the sky as the module orbits the Earth every 90 minutes. The maximum allowable power (cells plus battery) is 13kW on the sunlit portion of the orbit and 6kW in shadow.

The Environmental Control and Life Support System (ECLSS) maintains a sea-level atmospheric pressure of 14.7lb/sq in, maintaining a mixed gas atmosphere of oxygen and nitrogen. The nitrogen is manually controlled but oxygen is supplied by Russia's Elektron water electrolysis unit. This device collects and condenses moisture from humidity in the atmosphere and waste water which is broken down into hydrogen and oxygen using electrolysis. Hydrogen is vented to the vacuum of space and the oxygen is used to mix with the nitrogen to produce a breathable atmosphere. The system automatically maintains a constant output which, if it varies, is regulated by a change in current. This was new for Americans, who had never recycled atmospheric products in any of their spacecraft. The Russians had used the system in Mir, but with varying and sometimes unreliable results.

Water production is based on an average consumption of 20–25 litres per person per hour. The ECLSS also controls circulating fans, heat exchangers, fire detection equipment, and a gas analyser which detects pollutants. Thermal control is maintained by means of an ethylene-glycol and water mixture through two loops, one of which is a back-up. The loop's heat bearing fluids pass through 12 externally mounted radiator panels containing pipes filled with ammonia, transferring thermal energy to heat pipes and thence to space through radiation.

STS-88/ISS 2A Unity

Endeavour
4 December 1998, 8.36am GMT
Cdr: Robert D. Cabana (3)
Pilot: Frederick W. Sturckow
MS1: Jerry L. Ross (5)
MS2: Nancy J. Currie (2)
MS3: James H. Newman (2)
MS4: Sergei Krikalev (1) (Russia)

The second ISS element lifted into space was the first of three nodes. Called Unity, it was the first piece of US-built hardware lifted to the nascent station but the node was one of the most important components of the ISS because it had unique responsibilities. When NASA downsized the station from four primary US modules to two in an effort to cut costs it needed somewhere to put all the systems equipment essential to running the facility. So the concept of the node emerged, places where engineering and systems equipment could be located while also doubling as spacers between modules and other structures. Three nodes were built, one assembled by Boeing in the US (Unity) and two slightly bigger nodes (2 and 3) manufactured by agreement with the European Space Agency.

The launch of Node 1 and its two Pressurised Mating Adapters took place when Shuttle *Endeavour* lifted a crew of six from the Kennedy Space Center to an initial orbit of 175 x 87.2nm.

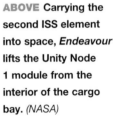

ABOVE Carrying the second ISS element into space, *Endeavour* lifts the Unity Node 1 module from the interior of the cargo bay. (NASA)

LEFT Installing Unity was a delicate job, the inertia of the 25,000lb structure controlled by precision movement using the Shuttle's Canadian-built manipulator. (NASA)

LEFT PMA-1 attached to Unity is gently lowered to the Docking Module on the Shuttle *Endeavour*. (NASA)

Through a series of manoeuvres *Endeavour* narrowed the gap to Zarya, 208nm above the Earth. On the second day after launch the Shuttle's remote manipulator arm grappled Node 1 and its two attached Pressurised Mating Adapters and lifted it from its cradle, moved it through 90 degrees and connected it to the Orbiter docking system at the forward end of the payload bay at a mission elapsed time of 39hr 9min.

Several additional rendezvous manoeuvres

ABOVE A fish-eye lens distorts the view of the Unity and its two PMAs attached to the Shuttle Docking Module and the Shuttle manipulator arm grasping Zarya. *(NASA)*

RIGHT Astronauts James Newman and Robert Cabana check out the Unity module. *(NASA)*

RIGHT Inside Zarya, astronauts reconfigure the equipment from a launch condition to an operational environment. *(NASA)*

were performed before *Endeavour* was at Zarya, about 60hrs 47min into the mission. The Shuttle manipulator arm was raised from its stowed position and at 63hrs 12min attached itself to the grapple fixture on Zarya and at 65hr 31min into the mission *Endeavour*'s thrusters were blipped to dock Unity with the Russian module. In Houston it was 8.07pm on 6 December and in Moscow it was 5.07am on 7 December. The first connection between two elements of the International Space Station had been achieved, creating a structure 76ft (23.2m) long and weighing over 31 tonnes. But a lot of work and two spacewalks were needed before the first human occupants could move from *Endeavour* to Unity and on into Zarya.

Jerry Ross and Jim Newman performed a 7hr 21min EVA on 7 December by leaving through the airlock beneath the docking interface with PMA-2 and hooking up 40 connections and cables. Activation sent power surging into Unity at an elapsed time of 91hr 13min. During the spacewalk they inspected the exterior of both Unity and Zarya and examined an antenna that had failed to deploy after launch. The next day the Shuttle's thrusters fired 11 times over a period of 21min 47sec to raise the combined orbit of the docked vehicles to 215 x 211nm. On 9 December a second EVA lasted 7hr 2min during which Ross and Newman removed restraint pins from the Common Berthing Mechanism on Unity, installed covers on two data relay buses, freed a back-up antenna on Zarya, and installed two S-band antennas on Unity. Only now could they prepare to open up Unity and float on through from the Shuttle's docking module.

Cabana and Krikalev floated into Node 1, side by side to symbolise partnership, at 155hr 18min elapsed time. In Houston it was 1.54pm on 10 December. Just 1hr 20min later, after bear hugs and a handshake, they opened the hatch into Zarya and floated across. In Houston ISS manager Randy Brinkley waxed unusually lyrical while Boeing Zarya manager Virginia Barnes summed it up, affirming that 'what happened today...signified not only tremendous engineering but also an emotional journey.' The next day, 11 December, the crew transferred 1,200lb (544kg) of equipment from *Endeavour* into Unity, stowing 335lb (152kg) of equipment from Zarya to the Shuttle. Turning

out the lights after 28hr 32min aboard the first two elements of the ISS, the crew withdrew into *Endeavour* and prepared for a final EVA.

The final spacewalk took place on 12 December when Ross and Newman freed an antenna on Zarya, leaving a tool strapped close by for future spacewalkers to use should they need it, and discovered that an experiment tray on the outside of the Russian module had a loose door flapping open. Moving more than 70ft (21m) above the Shuttle to the far aft end of Zarya, they marvelled at the view below them as they installed a new handrail that could be attached there before launch. Before returning through the airlock into *Endeavour* Ross tested a small emergency backpack called a SAFER (Simplified Aid For EVA Rescue) equipped with tiny nitrogen jets as thrusters, a device first tested in 1997 aboard Mir but which on that occasion had failed. The 6hr 59min EVA brought to an end the operational start to an assembly process that would last more than 12 years.

With PMA-2 depressurised, on 13 December the Shuttle and its docking module slipped away from the docked station at an elapsed time of 227hr 49min and began a fly-around inspection of the station. They had been docked for 6 days and 18 hours. After a final burn to start *Endeavour* drifting away from the ISS, preparations for re-entry began in earnest. The following day the crew released a small US Air Force satellite, MightySat, that had been carried in the payload bay, and a small satellite for Argentina. *Endeavour* landed back at the Kennedy Space Center on 15 December at an elapsed time of 11 days, 19 hours and 18 minutes.

ABOVE LEFT Jerry Ross conducts work outside Unity fixing cables and attaching conduits between Node 1 and Zarya. *(NASA)*

ABOVE Unity (foreground) attached to Zarya as viewed by a space walking astronaut. *(NASA)*

BELOW As *Endeavour* moved away its crew viewed the first unified building block for the ISS – Zarya and Unity. *(NASA)*

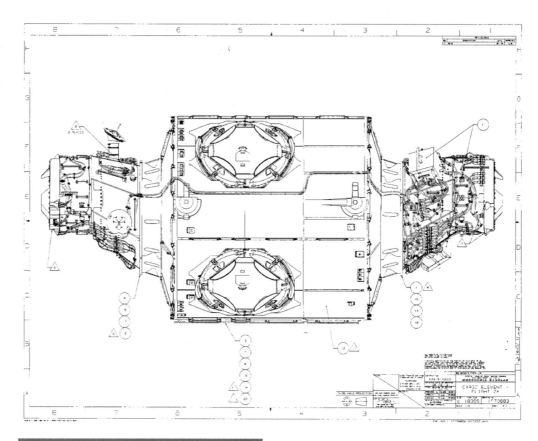

Anatomy

Node 1: Unity module/PMA-1 & 2

Length	18ft (5.5m) and 34ft (10.4m) including 2 PMAs
Diameter	15ft (4.6m)
Weight	25,600lb (11,600kg)
Launched	15 December 1998
Launch vehicle	Shuttle *Endeavour* STS-88
Launch site	Kennedy Space Center

Unity has six docking ports, one axial port at each end on the half-cone endplates and four radial ports placed at 90-degree intervals around the cylindrical structure. The six ports each have a Common Berthing Mechanism (CBM), a standard design of access and hatch used to link together all the non-Russian pressure modules throughout the ISS. They comprise two rings, an active CBM (ACBM) and a passive CBM (PCBM), performing the relevant functions their names imply and when connected together the two halves of the CBM form a pressure-tight seal. Each CBM incorporates a square hatch which when opened provides a 50-inch (1.27m) clearance for moving equipment between modules. The six docking ports on Unity are of the ACBM (active) type.

Unity was launched with forward and aft CBMs, each supporting a conical Pressurised Mating Adapter (PMA) about 8ft (2.44m) long. Each PMA comprised a truncated ring-stiffened shell structure machined from 2,219 aluminium roll-ring forgings welded together. With a 28-in (71cm) axial offset, all three PMAs used by the ISS are identical, acting as an interface between vehicles to which each was docked. Each PMA was an independent segment capable of being removed from Unity but they were preassembled with Unity for launch and on-orbit connection to other elements. As such, they constituted adapter units with a hatch at each end permitting pressure equalisation checks before opening the CBM into Unity. PMA-1 weighed 3,504lb (1,560kg) and joined Unity to Zarya at the aft end. Weighing 3,033lb (1,375kg), PMA-2 joined Unity to the Orbiter at the forward end, initially enabling Unity to receive Shuttle visits.

Each PMA carried hybrid computer multiplexer-demultiplexer units mounted externally. Pressurised internally and with handholds on the inside, initially they provided internal passageways into and out of the Unity node. PMA-1 would remain as the connection between the US and Russian sections for the life of the ISS but at various times during the assembly sequence PMA-2 would be moved to different locations, eventually taking up permanent residence at the top of the Harmony module (Node 2). PMA-3 would be delivered by STS-92 in October 2000 but each would provide primary docking ports for the Shuttle.

During the STS-116 mission of December 2006, PMA-2 would be fitted with the Station–Shuttle Power Transfer System (SSPTS), which served to provide power to the Orbiter while it was docked to the station. Incorporating a Power Transfer Unit (PTU) which replaced the Assembly Power Converter (APCU), it provided the ability to convert the 120 volt DC power from the ISS solar arrays to the Shuttle's 28 volt DC main bus. Up to 8kW could be fed from the ISS to the Shuttle, relieving demand from the Orbiter's fuel cells and adding up to four days docked at the ISS. The first operational use of this system came with the STS-118 mission in August 2007 but only *Discovery* and *Endeavour* were so modified, *Atlantis* being constrained to shorter visits.

An Androgynous Peripheral Assembly System (APAS) was attached to Zarya to connect it to PMA-1. An APAS consists of a structural ring, a moveable ring, various alignment guides, latches, hooks, and fixtures to mate with a copy of itself, each being either active or passive, hence 'androgynous'. A passive attach system was used on the forward end of PMA-2 to mate it to the Shuttle-docking module in the forward area of the cargo bay. During mating, the active half of the APAS capture ring is extended outwards from the structural ring towards the passive half and captures it, while an attenuation mechanism damps out relative movement between the two structures. With the two vehicles aligned, the capture ring is retracted inside the structural ring, whereupon 24 hooks lock the connection together to form an airtight seal. Two APASs were carried, one attached to each PMA.

Unity is equipped with four International Standard Payload Racks, or ISPRs, located at

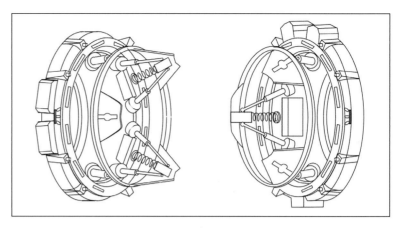

90-degree intervals around the periphery of the docking port that connects it to the Russian sector of the station, but it also doubles as an engine room with systems and equipment essential to the early integration and operational use of the evolving station. Fabricated from aluminium, Unity carried electrical power initially supplied by Zarya and distributed to other modules when they were attached, providing fault protection to individual branch lines. Items

ABOVE The APAS docking system used to connect the PMA to Zarya. *(NASA)*

LEFT The 'active' side of the APAS unit, shown here mounted to the Shuttle Docking module. *(NASA)*

LEFT The 'passive' side of the APAS unit attached to PMA-2 which would be used for docking to the Shuttle, PMA-1 being the permanent fixture between Unity and Zarya. *(NASA)*

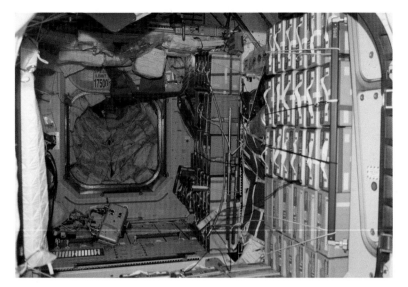

ABOVE Unity was launched with equipment stowed against the walls that would be unstowed in space and deployed in permanent locations. *(NASA)*

so attached included fans for air circulation, internal lighting, emergency egress battery chargers, communication systems, heaters, and facilities for future equipment installed on later visits. Power was supplied to Unity via cables routed through PMA-1, including umbilicals for secondary power, one for a data bus and three for power transfer. Six of the power umbilicals were attached during EVA operations conducted by the STS-88 crew of which two were for primary power and four for truss segments installed on later missions. Node 1 computers worked in unison with Zarya's computers to provide all ISS command and data handling services routed through three buses and a further three connected on later missions.

Thermal control was initially passive and then was provided by the Early External Active Thermal Control System (EEATCS) and finally by the External Active Thermal Control System (EATCS). Passive control involved shell-mounted patch heaters and multilayer insulation with Zarya supplying the electrical power for heaters. The EEATCS was installed with the P6 solar array package carried to the ISS by *Endeavour* on STS-97 in November 2000. These transferred heat using liquid ammonia in two identical loops operating at 35–41°F (2–5°C), each loop connected to an interface heat exchanger. The EATCS was activated after the assembly of the P3 and P4 trusses and associated solar arrays brought to the ISS by *Atlantis* on STS-115 in September 2006. The Internal Active Thermal Control System (IATCS) comprised the low, moderate and high-temperature thermal transport

loops. Because Unity was launched dry, these loops previously filled with nitrogen would only take on water when that was delivered by the Multi-Purpose Logistics Module (MPLM) Leonardo prior to the US laboratory module Destiny which brought the Fluid Systems servicer.

The Environmental Control & Life Support System (ECLSS) was designed to maintain a habitable environment and carried inter-module ventilation equipment, pressure sensors, and sundry ECLSS support equipment. The Atmosphere Control and Resupply system monitored the atmosphere and total pressure, measuring oxygen partial pressure, providing nitrogen and oxygen to the atmosphere and equalising pressure between adjacent modules. The Atmosphere Revitalisation section provided oxygen regeneration and removal of carbon dioxide, monitoring trace contaminants and hazardous atmospheres. Temperature and Humidity Control was responsible for removing moisture and heat from the air by circulating and ventilating the atmosphere and removing microbial airborne contaminants through filters, inhibiting 99.97 per cent of particles 0.3 microns or larger. The Fire Detection and Suppression system includes smoke detectors, a caution and warning panel, gas masks, and oxygen bottles and portable fire extinguishers. Unity carried 50,000 mechanical components, with 216 fluid and gas lines plus 121 internal and external cables running 35,000ft (10,668m) of electrical wiring.

Unity carried two multiplexer/demultiplexers running application software and process information, functions which are not normally carried in the multi-channel, simultaneous message transmission and de-channelling capabilities provided by such systems. Data and commands are exchanged via Mil-Std 1553B buses with an Intel 80386SX chip forming the base for the main processing card. The computers provide early command and control of Node 1 and were used for electrical functions and systems management including that of certain photo-voltaic control units. The computers are mounted on the exterior of PMA-1 on chassis platforms designed to protect them from debris and radiation. The computer cases are designed to slowly leak and equalise with their exterior environment, allowing them to be moved in and out of Unity if necessary.

STS-96/ISS 2A.1

Discovery
27 May 1999, 10.49am GMT
Cdr: Kent Rominger (3)
Pilot: Rick D. Husband
MS1: Tamara E. Jernigan (4)
MS2: Ellen Ochoa (2)
MS3: Daniel T. Barry (1)
MS4: Julie Payette (Canada)
MS5: Valery Tokarev (Russia)

The second flight to the ISS included a Russian cosmonaut and a Canadian astronaut on a logistics and resupply flight to keep Zarya and Unity up and running until the much delayed Zvezda could be launched. More than 3,600lb (1,630kg) of equipment in 750 separate items was uplifted by *Discovery*. Most of this equipment was housed within a 16,072lb (7,290kg) pressurised Spacehab double module carried at the rear of the cargo bay, access to which was gained by the astronauts along a tubular tunnel stretching from the Orbiter middeck the length of the cargo bay. The Spacehab double module is a 17ft (5.2m) long structure, 14ft (4.3m) wide and with a height of 11.2ft (3.4m). It had an internal volume of 1,100ft^3 (31.15m^3) capable of carrying 61 lockers or experiment racks, each 2ft^3 (0.057m^3) in volume. Developed for Phase I Mir missions, the double module would be used periodically during ISS build-up. In addition, mounted in the cargo bay above the access tunnel was a flatbed pallet and keel yoke assembly known as the Integrated Cargo Carrier (ICC).

Hail damage to the external tank while the Shuttle was on the pad caused a seven-day delay in the flight. Launch came at 6.50am local time on 27 May 1999, and *Discovery* was placed in an 11 x 183nm orbit, docking to PMA-2 on the forward end of Unity at an elapsed time of 41hr 48min. Just over two hours later the hatch to PMA-2 was open and the crew began to move inside Unity and Zarya, opening it up after a period of more than five months unoccupied. The single scheduled EVA of the mission began after two days and 16 hours of elapsed time from the Shuttle airlock, and, during a spacewalk lasting 7hr

55min, Jernigan and Barry transferred two folded cranes to the exterior of the station, set up new foot restraints, and installed tool bags and handholds for future assembly work. In all, 661lb (300kg) of cargo was transferred to the outside of the station.

About 13 hours after the conclusion of the EVA the Spacehab storage module was activated and the lengthy job of transferring all the equipment into the ISS began. A total of 2,881lb (1,307kg) in 98 items of dry cargo plus 686lb (311kg) of water with 197lb (89kg)

of cargo in 18 items moved into *Discovery* to bring home. After several days of activity, moving equipment, conducting experiments, and reconfiguring Unity and Zarya for the next phase, the Shuttle's thrusters were used to re-boost the altitude of the station. After a total of 5 days, 18 hours and 15 minutes docked together, *Discovery* separated from PMA-2 and drifted away, returning to Earth after 9 days, 19 hours and 13 minutes in space.

Anatomy

Integrated Cargo Carrier

Weighing 3,050lb (1,383kg) and constructed of aluminium, the ICC is 13ft (4m) wide across the bay and 8ft (2.44m) long with a depth of 10in (25cm). It incorporates a keel yoke for securing it to the bottom of the fuselage middeck, the lower part of the Orbiter cargo bay. The ICC is never removed during flight and serves effectively as a flatbed pallet to which can be attached up to 8,000lb (3,630kg) of cargo. The ICC was a preferred location of ISS Orbital Replacement Units (ORUs) carried to locations outside the ISS on modules and truss assemblies. For its first flight aboard STS-96 it carried 190lb (86kg) of parts for the Russian Strela crane, stowed on PMA-2 during a spacewalk, and 400lb (181kg) of EVA tools and flight equipment to assist crewmembers during assembly of the ISS.

The ICC was flown on STS-96, STS-101, STS-102, STS-105, STS-106, STS-108, STS-114, STS-121, STS-116, STS-122, STS-126, STS-127, STS-128, STS-131, STS-132, and STS-135. On STS-122 the version flown was known as ICC-Lite, being 4ft (1.2m) long rather than 8ft (2.4m) due to the small size of its load. Two flights, STS-127 and STS-132, carried an ICC-VLD (Vertical Light Deployable) version which differed in having heating and electrical power connections and being removable from the cargo bay using the Shuttle's robotic arm (see STS-127 payload description). Its empty weight was 2,645lb (1,200kg). Seven LCC missions (STS-108, STS-114, STS-121, STS-126, STS-128, STS-131, and STS-135) employed the Light Multi-Purpose Experiment Support Structure Carrier (LMC) which was a lightweight (946lb/429kg) cross-beam version without the keel yoke.

STS-101/ISS 2A.2a

Atlantis
19 May 2000, 10.11am GMT
Cdr: James D. Halsell (4)
Pilot: Scott J. Horowitz (2)
MS1: Mary Ellen Weber (1)
MS2: Jeffrey N. Williams
MS3: James S. Voss (3)
MS4: Susan J. Helms (3)
MS5: Yuri Usachev (Russia)

The launch of *Atlantis* was delayed for more than a month by bad weather and technical problems and was a logistics and resupply mission similar to that flown by *Discovery* a year earlier, carrying a double Spacehab module and an Integrated Cargo Carrier. The two docked modules had been unattended in space for a year, a necessity driven by extensive delays to the Russian Zvezda module which should have flown early in 1999. Held back by financial and operational problems, Zvezda was vital as the next module in the assembly sequence. But less than two months prior to the launch of *Atlantis* the Russians had finally bid farewell to Mir and brought it back to destruction in the Earth's atmosphere. Now they could focus on the ISS and cosmonaut Yuri Usachev, a veteran of 376 days aboard Mir and six spacewalks, accompanied five Americans when *Atlantis* was launched in May 2000.

Atlantis docked to PMA-2 at 1day 18hr 33min. But before entering the ISS, Voss and Williams performed a 6hr 44min spacewalk to transfer 372lb (169kg) of equipment outside including additional parts for the Strela crane which was moved to its permanent stowage location, more handholds, and tools. They also needed to perform several tasks setting up the exterior for the next missions. After the EVA the crew opened up the ISS modules and began logistics transfer and 'housekeeping' duties. The Spacehab double module was opened and 2,657lb (1,205kg) of dry cargo plus 187lb (85kg) of water in four contingency containers was moved to the ISS, with 1,291lb (586kg) moved into *Atlantis*. Extensive repair work on four of the six Zarya batteries turned astronauts into troubleshooters but technical exchanges between the ISS, Houston, and Moscow

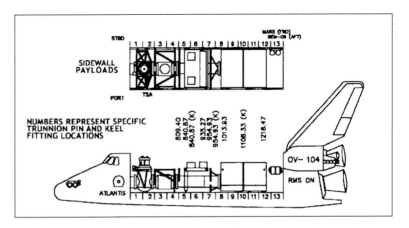

RIGHT Delays to the launch of Zvezda kept the
Zarya/Unity modules unmanned for a year but
STS-101 resumed operations in May 2000 with a
logistics supply mission using a double Spacehab
module. *(Spacehab)*

tracked down the problems and a decision was
made to replace two on the next mission.

A series of re-boost manoeuvres conducted
by the Shuttle raised the docked vehicles
to an orbit of 206.7 x 199.5nm, an altitude
increase of almost 10nm. The mission had been
extended by one day to accommodate the
battery troubleshooting and several other minor
tasks that built up on the timeline and when it
undocked at an elapsed time of 7 days 12hr
52min, *Atlantis* had been moored to the ISS for
5 days 18hr 19min. Engineers examining hours
of high-speed imagery of the launch discovered
a flash of light under the port wing indicative
of damage to thermal protection tiles on the
Shuttle, so a special type of re-entry was flown
to minimise heating in that area. Nevertheless,
when *Atlantis* landed it displayed several heavily
damaged tiles in other places and in one the
structure of the vehicle itself had almost burned
through, a nightmare that would re-visit NASA
less than three years later.

Zvezda/ISS 1R

12 July 2000, 4.56am GMT

Consisting of the Dos-8 core module
designed for Mir-2, the Zvezda spacecraft
is known as the Service Module, the first of
the Russian-owned modules launched to the
ISS. Zvezda means 'star' in Russian but, as
related earlier, this module was reluctant to
shine, with delays to its manufacture holding
back construction of the ISS by a year and then
delaying it further through postponed launch
dates. Launched more than two years behind
its original schedule, Zvezda was lifted into orbit
by Russia's powerful Proton-K launch vehicle.
Once separated from Proton, Zvezda used its
propulsion systems to adjust the orbit for a
rendezvous with the two on-orbit elements of
the ISS.

Using the Kurs rendezvous system,
Zvezda completed the approach phase and

ABOVE The Zvezda
service module during
manufacture and
assembly prior to
launch. *(RKK)*

LEFT Following long
delays, Russia's
Zvezda module is
launch by a Proton
rocket in July 2000,
the third ISS element.
(NASA)

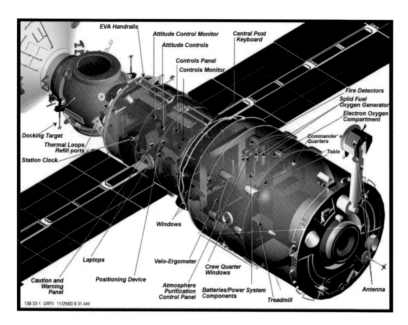

EVA Handrails
Attitude Control Monitor
Attitude Controls
Central Post
Keyboard
Controls Panel
Controls Monitor
Fire Detectors
Solid Fuel
Oxygen Generator
Electron Oxygen
Compartment
Commander's
Quarters
Table
Docking Target
Thermal Loops
Refill ports
Station Clock
Windows
Laptops
Velo-Ergometer
Crew Quarter
Windows
Caution and
Warning
Panel
Positioning Device
Atmosphere
Purification
Control Panel
Batteries/Power System
Components
Treadmill
Antenna

138-33-1 GRFX 11/29/00 9:31 AM

ABOVE Zvezda was the second major Russian element added to the station, greatly increasing its capacity and adding electrical power. *(NASA)*

then became the passive vehicle while Zarya accomplished the docking. The date was 26 July in Moscow and 25 July in the US. Now comprising three elements, the Zvezda–Zarya–Unity assembly had a total length of almost 120ft (36.5m) and a mass of 110,000lb (49,900kg). For six weeks the docked assembly would remain unattended, awaiting the arrival of *Atlantis* on its second visit to the ISS that year.

Before that, on 8 August 2000, the first unmanned cargo-tanker vehicle sent to the ISS docked to the aft port on Zvezda. Launched two days earlier, Progress M1-3 carried supplies to the station and was ready for the crew of *Atlantis* when it arrived a month later. M1-3 would remain docked to the aft Zvezda port for 84 days until de-orbited on November 1.

Zvezda

Length	43ft (13.1m)
Diameter	13.5ft (4.1m)
Pressurised volume	2,649ft^3 (75m^3)
Habitable volume	1,649ft^3 (46.7m^3)
Solar array span	97.5ft (29.7m)
Weight	42,000lb (19,050kg)
Launched	20 July 2000
Launch vehicle	Proton
Launch site	Baikonur

The Zvezda module had a long pedigree dating back to the 1980s and even the structural layout is nearly identical to the core module of the Mir space station. Zvezda comprised four primary sections: the Transfer Compartment, the Work Compartment, the Transfer Chamber, and the Assembly Compartment. The spherical Transfer Compartment is at the front and provides the main interface with Zarya attached at its main axial docking port. Unlike the Mir core module, Zvezda has only two additional docking ports in this compartment – one up and one down – and unlike Mir it has no Lyappa system for transferring modules. These ports were designed to receive Soyuz and Progress spacecraft but other small modules have been attached, including the Pirs docking compartment and the Mini-Research Module-2. Pirs is like the docking module attached to Mir and allows two astronauts to exit Zvezda without depressurisation. Zvezda was built to accommodate three crewmembers, or six in a transfer function.

RIGHT With a pedigree back to the 1980s, Zvezda owes much to the Mir core module design and is functionally divided into separate working areas. *(NASA)*

RIGHT Cosmonaut Yuri Usachev in Zvezda working a laptop. *(NASA)*

The Transfer Compartment consists of a small, spherical volume with a forward docking port to engage the aft end of Zarya, a zenith (upper) docking ring on top, and a nadir (lower) docking ring at the bottom. Aft of the Transfer Compartment, the cylindrical main Work Compartment is divided into a forward section 9.5ft (2.9m) in diameter and an aft section 13.5ft (4.1m) in diameter providing most of the living and work space in the Russian sector of the ISS. The forward section contains engineering panels, caution and warning displays, lighting control panels, maintenance equipment, and a body mass measurement device.

The aft section contains the environmental control equipment and waste management (toilet) systems together with two crew compartments for individual astronauts. Eventually, it was also equipped with a US Treadmill & Vibration Isolation System (TVIS) and the Russian Vela Ergometer for exercise and physical fitness monitoring. The Work Compartment also provides a kitchen area with a refrigerator-freezer and a table platform for securing meals in the weightless environment. Altogether, Zvezda has 13 optical ports including a 9in (23cm) diameter window in the Transfer Compartment for viewing docking operations, a large 16in (41cm) window in the Work Compartment, and a single window in each crew compartment, which double as sleep stations. Other windows are situated at various locations for viewing the Earth and other modules.

At the extreme aft end of Zvezda is the Transfer Chamber, essentially a short section through which crewmembers can pass from the Work Compartment to the aft docking port, the fourth with which Zvezda is equipped. Wrapped around this tunnel is the Assembly Compartment, comprising the aft end of the module, which contains the propellant tanks and rocket motors for orbit adjustment and attitude control. The Zvezda propulsion systems use the same N2O4/UDMH propellant combination as Zarya. Mounted on the aft face of the module, the two main engines each have a thrust of 661lb (300kg) and can be fired individually or in pairs for orbit adjustment and manoeuvring. In addition, there are two rings of 16 redundant thrusters, each delivering a force

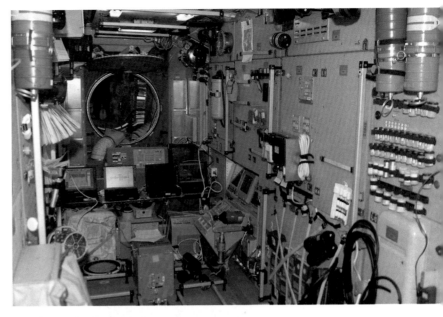

ABOVE **A view looking forward toward the spherical Zvezda Transfer Compartment.** (NASA)

LEFT **As assembly progressed and logistical support increased, Zvezda became increasingly crowded. Krikalev floats through the main work compartment.** (NASA)

LEFT **Sergei Krikalev watches** Atlantis **through one of the windows in Zvezda.** (NASA)

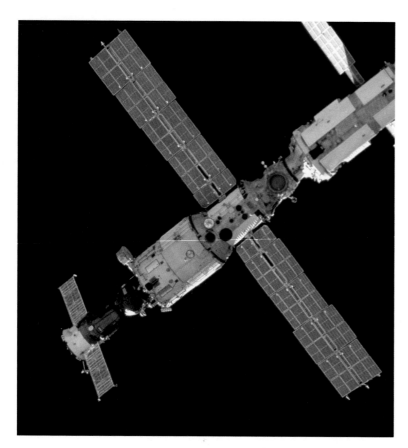

The European Space Agency provided the Data Management Systems, the brains of Zvezda, which were responsible for controlling all module functions as well as those of the other station elements added, until those functions were taken over by the US Destiny module. Constituting the first European equipment sent as part of the ISS, the DMS was put together by an industrial consortium led by Daimler-Chrysler of Bremen, Germany. ESA supplied this equipment to the Russians in return for two flight-rated docking systems that would be used later with Europe's Automated Transfer Vehicle (ATV).

STS-106/ISS 2A.2b

Atlantis
8 September 2000, 12.46pm GMT
Cdr: Terence W. Wilcott (3)
Pilot: Scott D. Altman (1)
MS1: Edward T. Lu (1)
MS2: Richard A. Mastracchio
MS3: Daniel C. Burbank
MS4: Yuri Malenchenko (Russia)
MS5: Boris Morukov (Russia)

of 29lb (13kg) for attitude control. Propellant is stored in four 52.8-gallon (240-litre) tanks located in this unpressurised section, holding a total of 1,896lb (860kg) of fuel and oxidiser. Expulsion of the hypergolic propellant utilises a nitrogen pressurisation system for flow management to the motors and thrusters.

The Elektron oxygen generation system became the standard unit for supplying all compartments with oxygen but frequent problems resulted in the use of a Solid Fuel Oxygen Generator, or SFOG. More commonly known as oxygen candles, they comprise a cylindrical generator with a mixture of sodium chlorate and iron powder smouldering at 1,112°F (600°C). This produces sodium chloride, iron oxide and about 6.5 man-hours of oxygen per kilogram of stored fuel. Exhaled carbon dioxide from the crew would build up to dangerous levels if not removed and Zvezda is equipped with the Vozdukh system, which uses regenerating absorbers to scrub the air and keep it clean. Water condensates are extracted from the air by the SKV system which is then processed by the Russian Condensate Water Processor, or SRV-K.

With two of the three assembled modules built in Russia, it was only natural that the next logistics and servicing mission to the ISS should carry two cosmonauts in its seven-man crew. *Atlantis* carried a double Spacehab module and the ICC stacked with supplies, its primary objective being to change Zvezda from the launch to the flight configuration by installing six ground repressurisation inlet caps, remove the fire extinguisher launch restraint bolts, and activate the gas masks for the first permanent resident crew (known as Expedition 1). For the first time the crew was able to offload 1,300lb (590kg) of supplies from the Progress M1-3 cargo vehicle, which had been launched on 6 August. The crew removed the TORU docking unit on Zvezda and the aft docking probe from Zarya, thus facilitating movement between the two modules. Two replacement and three new batteries were installed in Zarya along with voltage converters and other electrical equipment.

In all, 5,399lb (2,449kg) of stores were moved from Shuttle to the ISS – including 780lb

(354kg) of drinking water – with 948lb (430kg) returned to Earth in *Atlantis*. On a 6hr 14min spacewalk, nine data, communication, and power cables were hooked up between Zvezda and Zarya and a magnetometer installed on the outside which would be used as an equivalent Earth compass. Acoustic measurements were made in Zvezda and a toilet installed along with the treadmill exercise equipment. In four re-boost manoeuvres using Shuttle thrusters the ISS was raised to an orbit of 208.6nm x 203.8nm, increasing altitude by about 13nm. The Shuttle remained docked to the ISS for 7 days 21hr 41min, a one-day extension seeing all the scheduled tasks accomplished as planned. *Atlantis* landed after a flight lasting 11 days 19hr 11min.

STS-92/ISS 3A Z1 Integrated Truss Segment

Discovery
11 October 2000, 11.17pm GMT
Cdr: Brian Duffy (3)
Pilot: Pamela A. Melroy
MS1: Leroy Chiao (2)
MS2: William S. McArthur (2)
MS3: Peter J. K. Wisoff (3)
MS4: Michael E. Lopez-Alegria (1)
MS5: Koichi Wakata (1) (Japan)

The fifth Shuttle flight in support of the ISS, STS-92 was the first major assembly flight, carrying the Z1 Integrated Truss Segment which was to be attached to the top (zenith) port on Node 1 (Unity). In addition, *Discovery* carried

ABOVE Seen from *Atlantis* on STS-106 in September 2000, the two Russian modules provide all the electrical power for their own function as well as the Unity module (below). *(NASA)*

LEFT Location of attachment points between Zarya and Zvezda. And the connections made during the first space walk on the STS-106 mission in September 2000. *(NASA)*

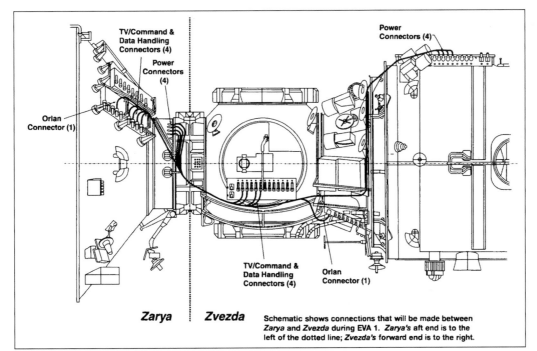

TV/Command & Data Handling Connectors (4)
Power Connectors (4)
Power Connectors (4)
Orlan Connector (1)
TV/Command & Data Handling Connectors (4)
Orlan Connector (1)

Zarya | **Zvezda**

Schematic shows connections that will be made between *Zarya* and *Zvezda* during EVA 1. *Zarya's* aft end is to the left of the dotted line; *Zvezda's* forward end is to the right.

Pressurised Mating Adapter 3 (PMA-3), the last such unit to be flown to the ISS, carried on a Spacelab logistics pallet inside the Shuttle cargo bay. It would be berthed at the lower (nadir) port on Node 1. Two EVA tool stowage devices were also carried up along with two DC-to-DC converters to be installed on the Z1 truss atop Node 1. This was a complex flight involving a lot of structural build-up, demanding four EVA trips outside through the Orbiter's airlock.

Carrying a payload of 35,250lb (15,990kg), *Discovery* latched on to PMA-2 at 46hr 28min, where it remained for a total of 6 days 21hr 10min. The first task was to enter PMA-2 and conduct an air quality test, but only when the Z1 truss had been retrieved from its location in the cargo bay and positioned on to the Node 1 zenith port could work commence in the ISS. The Z1 truss box was grappled by the Shuttle's manipulator arm at an elapsed time of 88hr 30min and slowly manoeuvred on to the zenith docking port on Unity, the arm releasing it 3hr 8min later. The crew then entered Unity and installed grounding straps along with other equipment necessary to support the upcoming spacewalks. Crowded with tasks, the mission supported four spacewalks on four successive days, the highest level of workload carried by

any crew thus far. Spacewalkers alternated so that each pair would get a day off between their two EVAs.

The first EVA took Chiao and McArthur on a 6hr 28min spacewalk on 15 October connecting power cables to heaters and connecting conduits on the Z1 truss, relocating two antenna assemblies, and placing a toolbox for future assembly work. Also conducted during this EVA was the transfer of PMA-3 from the Shuttle cargo bay across to the Earth-facing nadir port on Unity. This would be the port to which the next Shuttle would dock, rather than PMA-2, so as to leave an area around Unity free to deploy the massive and sizeable P6 truss and solar array assembly. Next day, Lopez-Alegria and Wisoff installed PMA-3 on the Unity nadir port (facing Earth) and continued work on the Z1 truss, preparing it for installation of the P6 solar arrays in an EVA lasting 7hr 7min.

A day later, Chiao and McArthur conducted a 6hr 48min excursion primarily working to fix and set up the DC-to-DC converter assemblies on top of Z1, also reconfiguring power and data umbilical cables to support docking of *Endeavour* to PMA-3 in November. The final EVA took place on 18 October with Lopez-

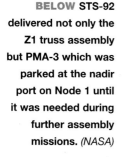

BELOW STS-92 delivered not only the Z1 truss assembly but PMA-3 which was parked at the nadir port on Node 1 until it was needed during further assembly missions. *(NASA)*

LEFT **In October 2000**
***Discovery* brought the
Z1 truss assembly and
parked it on top of the
Unity module, seen
here front on.** *(NASA)*

Alegria and Wisoff removing the grapple fixture
from the Z1 truss, setting up a utility tray,
recycling and opening the manual berthing
mechanism latches on Z1, and conducting
further tests on the emergency SAFER system
designed to provide limited propulsion for a
free-flying astronaut. In a 6hr 56min spacewalk
they concluded four EVAs on four consecutive
days lasting 26hr 38min, setting up the ISS for
the next two Shuttle visits and establishing a
trend as assembly and installation of the ISS
collected momentum.

Between EVAs and after the last spacewalk,
the Shuttle conducted three re-boost
manoeuvres that left the docked assembly in
an orbit of 214 x 202nm, a standard activity as
the tenuous traces of Earth's outer atmosphere
sought to slow the ISS and pull it Earthwards.
Transfers and the movement of cargo, freight,
and stores shifted a total of 21,998lb (9,978kg)
to the ISS including 1,098lb (498kg) of sundry
equipment. Undocking took place after 8 days

15hr 51min in anticipation of landing two days
later, but three attempts to land at the Kennedy
Space Center were cancelled before the de-orbit
burn due to bad weather in Florida, so *Discovery*
landed at the back-up site, Edwards Air Force
Base, California, at an elapsed time of 12 days
21hr 43min.

This was the last time elements of the ISS
would be left unmanned. Now it was time
to start an unending period of permanent
habitation – Phase 3 of ISS assembly –
expanding from an initial complement of three
to a six-person crew involving scientists and
engineers from all the station partner countries.

Anatomy

Z1 Integrated Truss Assembly

Z1 would support the early use of solar arrays
that, over time, would be moved across the
evolving ISS until finally positioned at the
extremities of a long truss boom composed of
13 separate sections brought up on individual

flights. Z1 weighed 21,998lb (9,978kg) and would temporarily reside at the upper node on Unity but it would also serve as a platform for systems that would be moved about. Primarily, Z1 would provide a mounting base for the P6 truss assembly and its associated solar array. The 12 primary truss structures were given letter designations 'P' for port (left) and 'S' for starboard (right). Those positions were referenced to left and right sides of the ISS when looking down from the front of the station to the rear, where the two Russian modules Zarya and Zvezda reside.

An important function of the Z1 truss is as a command and tracking subsystem, with an S-band communication system consisting of two redundant sections each with three replaceable units: a baseband processor and low-gain and high-gain RF directional antennas. It serves to amplify and filter radio signals, with the high-gain antenna supporting the high-rate data link. Several functions would not be available until Mission 5A launched in February 2001 and some subsystems installed on Z1 would only become active when additional equipment arrived on later flights, another example being the transponder for the Tracking

and Data Relay Satellite System (TDRSS) on Mission 4A in November 2000.

The TDRSS is a constellation of communication satellites in geostationary orbit that receive and transmit data from and to a wide range of satellites in low Earth orbit, including the Shuttle and the ISS. Routing signals in this way saves money otherwise spent on ground tracking stations located around the world. The Ku-band system on Z1 would become the primary return link for ISS video and payload data transmitted in digital form at a downlink rate of 50-Mbps with up to four video signals, or 43-Mbps high-data rate. Usually, transmissions are configured from the ground through the S-band link and can be full-motion video or stop-action video, which consists of skipping video frames. The video baseband signal processor has four processor units feeding to an erectable, steerable antenna.

Z1 carries several electrical power distribution system components to provide a connection path from the P6 solar array brought up on Mission 4A and interlinks with other arrays brought up later. Power control modules, data connections and a cold plate thermal system are provided to protect and isolate secondary lines. Power Control Units are essential for controlling the voltage between the space plasma and the ISS structure to within 40 volts of the plasma potential. Mounted to Z1, they emit electrons through a self-generated plasma whenever the ISS is in sunlight, without which parts of the ISS structure could reach potentials of -150 volts. DC-to-DC conversion units were attached to the Z1 truss during the first spacewalk.

Not activated until the arrival of the P6 solar arrays with Mission 4A, the units convert 115–173 volts DC primary to 123–126 volts DC secondary power. Two patch panels on the side of Z1 allow the source of the power input to be changed, and three interchangeable connectors route energy to the Russian segment to keep the system alive until switched to the P6 array. Z1 also provided several spacewalking aids, including two EVA tool stowage devices, 22 worksite interface sockets, a mobile grapple fixture, and numerous handholds and restraints for fixing to a wide range of locations.

One of the most important packages aboard

BELOW The Z1 truss would provide vital attitude control capability as well as power transfer capabilities to the growth of ISS assembly. (NASA)

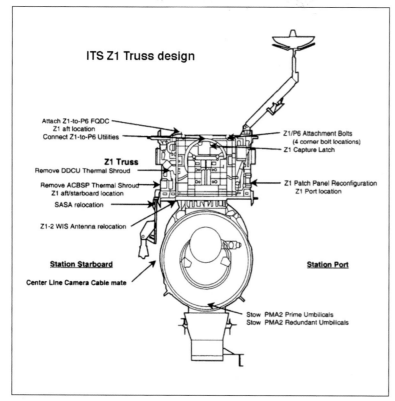

ITS Z1 Truss design

Attach Z1-to-P6 FQDC
Z1 aft location
Connect Z1-to-P6 Utilities

Z1 Truss
Remove DDCU Thermal Shroud

Remove ACBSP Thermal Shroud
Z1 aft/starboard location

SASA relocation

Z1-2 WIS Antenna relocation

Station Starboard

Center Line Camera Cable mate

Z1/P6 Attachment Bolts
(4 corner bolt locations)
Z1 Capture Latch

Z1 Patch Panel Reconfiguration
Z1 Port location

Station Port

Stow PMA2 Prime Umbilicals
Stow PMA2 Redundant Umbilicals

the Z1 truss assembly was the complement of four Control Moment Gyro (CMG) units. They would form a fundamental core of the station's pointing and attitude control mechanisms, storing momentum in massive spinning wheels placed orthogonally at 90-degree intervals to each other to control roll, pitch, and yaw. The only alternative to CMG technology would be the almost continuous use of thrusters consuming propellant at an unsustainable rate. CMGs are essentially spinning gyroscopes that if slowed down or speeded up will dump or increase energy in their resistance to deflection. Each CMG consists of a large, flat wheel spinning at 6,600rpm, developing an angular momentum of 3,500ft-lb/sec about its spin axis. The four CMGs therefore have momentum storage of 14,000ft-lb/sec, which is the scalar sum of the individual wheel moments. This wheel is mounted in a two-degree-of-freedom gimbal system that can point the wheel in any direction.

To maintain the ISS in the desired attitude, the CMG system must cancel, or absorb, the momentum generated by the disturbance torques acting on the station. At least two CMGs are needed for attitude control, the system generating an output reaction torque by inertially changing the direction of its wheel momentum. The output torque has two components, one proportional to the rate of change of the CMG gimbals and a second proportional to the inertial body rate of the ISS as sensed at the base of the CMG. Because the momentum along the direction of the spin axis is fixed, the output torque is constrained to lie within the plane of the wheel, which is why one CMG cannot provide three-axis torque necessary for full attitude control. Each wheel unit contains a thermostatically controlled heater to keep thermal limits between a range of between -42°F and -35°F (-41°C and -37°C), utilising heaters rated at 120 watts. The CMG units would not be activated before Mission 5A.

The Z1 truss also provided a manual berthing mechanism for temporary stowage of PMA-2, a device similar to the one used on the Spacelab pallet to mount it within the Shuttle's payload bay. Presently used as a Shuttle docking port, eventually PMA-2 would be moved around the station and this location would serve as a suitable berthing place. Z1 also supported a hinged cable tray that would be used to hook up power and data cables as well as coolant lines to the US laboratory module Destiny when that docked to Node 1 during the STS-98 mission. The Z1 structural shell was fabricated from aluminium 2219-T851 with trunnion and keel pin beams in INCO 718, with a solid plate being fixed beneath the CMG units to protect them from micrometeoroid impacts. The Ku-band antenna boom beam elements were fabricated from steel. Thermal control for the Z1 truss included additional elements of the EEATCS (see Node 1), including four accumulators that would charge the control system in orbit and maintain the system's operating pressure.

Anatomy

PMA-3

With a weight of 2,549lb (1,156kg), PMA-3 was the third and last of its type uplifted to the ISS and was essentially the same as the two lifted up attached to the Unity module. Like PMA-2, this most recent addition provided a docking interface for the Shuttle but it also included mechanical interfaces, EVA hardware and thermal conditioning equipment, an electrical power subsystem, and data handling cables. The command data handling system was based around a hard-line 1553 data bus between the Shuttle and Unity through x-connectors on the PMA-3 APAS (see Unity). This continued the routing from Unity across to the Shuttle, allowing interface units and Orbiter portable computer systems to talk on the ISS Orbiter buses.

Spacewalk support included prepositioned handrails and foot restraints with a grapple fixture for a remote manipulator arm to move the PMA-3 around, even during an EVA. Mechanical interfaces for PMA-3 include a Passive Common Berthing Mechanism and a Russian APAS, allowing Soyuz vehicles to dock. Thermal control is provided by ten 60-watt passive thermal control heaters and multilayer insulation. Shell temperature is maintained by electrical resistance heater circuits and the heater set-points can be adjusted in space or by ground control.

Chapter Four

ISS Phase Three – permanent habitation

Having learned how to work together on Earth, Americans, Europeans, Canadians and Japanese began to live together in space – inside a gradually evolving complex of modules and gantries the like of which had never been attempted before – a place where scientists and engineers could work on tasks together.

OPPOSITE The three Expedition 2 crewmembers and the seven STS-100 crewmembers gather together for a photo-shoot in Europe's Leonardo module, April 2001. *(NASA)*

As planned, the prime elements of the Russian segment of the International Space Station had been the first to be assembled, largely because the Russians already had the template for one of the modules as a prelude to the now defunct Mir-2. But the Russians had been loath to abandon Mir itself, and this played a major role in slowing the overall development of the ISS. Criticism of Russia's split role, managing both their own Mir facility and fulfilling their obligation as a partner to the ISS, ignited concerns among some US politicians. Outright opposition to the idea of a space station and all it implied for long-term funding demands had polarised efforts to get it cancelled and these delays fed into that agenda.

The flight of Discovery in May 1999 had been necessary to keep Zarya and Unity running without a permanent crew, and the launch of Zvezda more than a year later had been a welcome restart for an ambitious assembly sequence delayed by both political and technical problems. That, plus the flight of STS-92 and installation of the Z-1 truss assembly on the Unity node, provided welcome reassurance that the ISS was growing into a fully fledged research laboratory and that the facility could now begin expeditionary flights as a permanently manned orbiting research laboratory.

The preceding two years had been difficult on several levels. While astronauts and cosmonauts had learned how to work together, managers in participating countries had been frustrated by the countless delays and difficulties encountered by various procedures. Hardware intended for launch was held back by the delays to getting the Zvezda module launched, and costs increased as sensitive equipment and delicate systems were kept in top condition in holding facilities in Europe, Japan, Canada and the United States. Now it was time to begin living in what was still a construction site in space.

Soyuz TM-31/ISS 2R/ Expedition 1

31 October 2000, 7.52am GMT
Cdr: William B. Shepard (US)
Pilot: Sergei K. Krikalev
Flight Engineer: Yuri Gidzenko

On 31 October 2000, a Soyuz launch vehicle lifted off from Baikonur carrying the first crew to begin permanent habitation aboard the ISS. Commander of Expedition 1, NASA astronaut William B. Shepard, was accompanied by Russian cosmonauts Yuri Gidzenko and Sergei Krikalev. In its agreement with Russia the US insisted that a NASA astronaut would command the first mission to begin permanent occupation. The Russians had selected the highly experienced cosmonaut Anatoli Solovyov to fly but were reluctant to allow an astronaut who had never operated aboard a station to have command over a man with substantial flight time aboard Mir. A former Navy SEAL, Shepard had made three flights aboard the Shuttle for a total of less than three weeks in space. Solovyov had flown six flights and logged 651 days in space, conducting 16 EVAs totalling more than 82 hours. The less experienced Gidzenko replaced Solovyov, but even Krikalev had greater experience than Shepard, having had four flights, some aboard Mir, totalling a year in space.

Launched aboard Soyuz TM-31 with Gidzenko in command for the two days it took to reach the ISS, the official start to the permanent occupation of the ISS by Expedition 1 began with the docking at Zvezda's aft port at 9.21am (GMT) on

BELOW The first permanent crew delivered to the ISS included the Commander William B. Shepard from the US, Pilot Sergei K. Krikalev and Flight Engineer Yuri Gidzenko from Russia. *(NASA)*

2 November, a port vacated by Progress M1-3 a day earlier. Shortly after docking the crew entered Zvezda, switched on the lights and began preparing for their first meal aboard the ISS, whose radio call-sign was 'Alpha' at the request of Shepard. An early task was for Gidzenko and Krikalev to don Orlan space suits and, sealing themselves inside the spherical Forward Transfer Compartment, to relocate the docking probe used to connect to Zarya from the forward to the nadir (Earth-facing) docking port.

Life aboard the ISS was regulated around GMT (Greenwich Mean Time or Universal Time, UTC) with wakeup at 5.00am. First up was Progress M1-4 which arrived at the Zvezda nadir port on 18 November, a remotely controlled manual operation when the automatic Kurs system failed. It was a month before Unity could be used for communication with Earth, utilising TDRSS relay equipment brought up by their first visitors from STS-97. Progress M1-4 had undocked on 1 December to make space for *Endeavour* and re-docked on 26 December after they left, finally departing to a fiery re-entry on 8 February 2001. Then Soyuz TM-31 was relocated from the aft to the nadir port on Zvezda, making room for Progress M-44 which docked at the aft port on 28 February.

Shepard, Gidzenko and Krikalev hosted two more Shuttle missions (STS-98 and STS-102) before handing over to the Expedition 2 crew

and returning to Earth aboard *Discovery* on 21 March 2001, by which time they had spent 136 days 17hr 9min at the station, a total time in space of 140 days 23hr 38min.

STS-97/ISS 4A P6 Truss & Solar Array

Endeavour
1 December 2000, 3.06am GMT
Cdr: Brent W. Jett (2)
Pilot: Michael J. Bloomfield (1)
MS1: Joseph R. Tanner (3)
MS2: Marc Garneau (2) (Canada)
MS3: Carlos I. Noriega (1)

By the time *Endeavour* was launched as the last Shuttle mission in 2000, projected completion of the ISS had already drifted out

ABOVE The Soyuz TM-31 transport that carried the Expedition 1 crew to the ISS in October 2000 is rolled out to the launch pad. *(NASA)*

LEFT The visiting STS-97 crew meets for a photo-shoot with the incumbent Expedition 1 ISS crew. *(NASA)*

17hr 4min. Just 2hr 16min later the Shuttle's remote manipulator arm grappled the P6 array from its stowed position in the cargo bay and extended it above the Shuttle where it would remain overnight for thermal balance. While this was going on the outer hatch to PMA-3 was opened and some items placed inside ready for transfer to the ISS, but not for several days would *Endeavour*'s crew join Expedition 1. This was so that the Shuttle could operate at a reduced 10.2lb/in^2 atmospheric pressure, easing the time required for spacewalkers to condition to the pure oxygen of their suits. The ISS operated at 14.7lb/in^2.

The first spacewalk began at 2 days 15hr 30min, when Tanner and Noriega left *Discovery* to move across ready to attach the P6 assembly to the Z1 truss deposited atop Unity on the previous Shuttle flight. Still clutching the 35,000lb (15,875kg) truss and solar array assembly, the manipulator arm was commanded by Garneau to bring it across to Z1, where it was positioned for the crew to secure. Activating capture latches on the Z1 truss, Noriega brought the P6 array into fine alignment, securing it with 127 turns of the capture latch assembly. Both crewmembers secured bolts at each of the four corners of the P6 box before releasing the latch to allow loads to go through the securing bolts. Garneau released the arm from the grapple fixture at 1hr 41min into the EVA and the spacewalkers spent the remainder of their 7hr 33min time outside setting up the P6 for operations and releasing the arrays. But difficulties with the deployment of the array's PV blanket caused delays and problems that kept mission control scurrying for ways to correct faulty unfurling of both blankets.

A second EVA lasting 6hr 37min was conducted two days later when Tanner and Noriega configured the electrical connections to the P6 array, positioned the S-band antenna for ISS use, and connected cables and conduits for the arrival of the US Destiny module. A third spacewalk two days after the second lasted 5hr 10min and completed repairs to the tension line on the starboard P6 array. They also installed a camera cable outside Unity. Spacewalking over, the *Endeavour* crew could now equalise pressure between the two structures and move in to the ISS. For just over 25 hours the two

to 2006 from its original date of 2002. But the additional power resource that STS-97 carried to the station was a real start at providing a permanent home and work environment for its occupants. The prime objective of this mission was to deliver and install the P6 Truss & Solar Array Assembly and its associated equipment to the Z1 truss segment. P6 indicated that its ultimate location would be on the port (left) side of the station. PV (photovoltaic) Module P6 was the first of eight such arrays that would be lifted to the station, with a total payload mass of 42,804lb (19,416kg).

A day after *Endeavour* reached orbit Progress M1-4 was undocked from the ISS making space for the Shuttle to arrive. *Endeavour* docked with PMA-3, using it for the first time as the point of access to the ISS, at an elapsed time of 1 day

crews exchanged conversation, ate together, and transferred 1,457lb (661kg) of supplies to the ISS including 684lb (310kg) of water. *Endeavour* received 227lb (103kg) of redundant items from the station. When the Shuttle undocked at 8 days 16hr 7min, the two had been locked together for 6 days 23hr 1min. Mission duration for the Shuttle was 10 days 19hr 57min.

ANATOMY

P6 Truss & Solar Arrays

P6 consists of two identical Photovoltaic Array Assemblies (PVAAs), each of which contains two Solar Array Wing (SAW) assemblies (2B and 4B) connected to a mast folded into a Mast Container prior to deployment for a total weight of 185lb (84kg). A Beta Gimbal Assembly (BGA) allows the extended solar arrays to pivot around their mounting to track the passage of the sun as the ISS remains fixed in its orbital path around the Earth. Additional equipment would be used to transfer power from the rotating gimbal, to control the mast and mast rotation, to regulate the power output voltage, and to route it to integrated electrical assemblies.

The two solar arrays wings on the P6 module deploy in opposing directions to each other. Each SAW is made up of two solar panels, each folded into its own Solar Array Blanket Box (SABB) measuring 20in (51cm) in height and 15ft (4.57m) in length. When fully deployed each SAW extends 108.6ft (33.1m) and spans 38ft (11.6m) across, presenting a total span

of more than 240ft (73m) when deployed in opposing directions. Each solar wing is the largest deployed in space, weighing more than 2,400lb (1,089kg). Each wing is split into two blankets separated by the central telescoping mast that extends or retracts to form or fold the array wing. Each wing supports 16,400 silicon cells, each measuring 8cm^2. They are grouped into 82 active panels, each consisting of 200 cells, with 4,100 diodes. Each SAW is capable of delivering almost 33kW of DC power, thereby providing almost 66kW for the total power generation of the P6 array, which is more than enough for 30 average homes.

The BGA's most vital functions are to deploy and retract the SAW and to rotate it about its longitudinal axis. The system has a pointing accuracy of +/– 1 degree with positional

ABOVE Canadian astronaut Marc Garneau floats from unity to PMA-3 during the STS-97 flight that delivered the first (P6) independent solar arrays. *(CSA)*

LEFT The P6 array dwarfs the solar arrays on Zarya and Zvezda from its temporary position on top of the Z1 truss mounted to the zenith port on the Unity module. *(NASA)*

RIGHT To deploy, the
solar arrays unfold
from their accordion-
like box using a
spool system with
guides and runners to
maintain uniformity.
(NASA)

authority under the control of a three-phase, 200W DC stepper motor. It has a maximum rotational rate of +/– 200 degrees per minute and the motor has an operating torque of 380in-lb and a maximum stationary torque of 1,700in-lb. The BGA measures 8ft x 8ft x 2ft (2.44m x 2.44m x 0.61m) and provides a structural support for the Integrated Electronics Assembly (IEA). The Sequential Shunt Unit (SSU) is designed to coarsely regulate the solar power collected during periods of insolation, when the array collects power during sun-pointing periods. A sequence of 82 separate strings, or power lines, lead from the solar array to the SSU and shunting by controlling the amount of power transferred. The SSU weighs 185lb (84kg) and measures 32in x 20in x 12in (81cm x 51cm x 30cm).The Integrated Electronics Assembly (IEA) constitutes a box-shaped structure measuring 16ft (4.88m) on a side and weighing 17,000lb (7,711kg). It is designed to condition and store the electrical power collected by the SAW and used aboard the station. The IEA power system is divided into two independent channels and power from each array is fed directly into the appropriate Direct Current Switching Unit (DCSU) – a high-power, remotely controlled unit used for both primary and secondary power distribution and fault isolation – and is also responsible for primary power distribution to the ISS during periods of eclipse. Power from the DCSU is also distributed to the DC to DC Converter Unit (DDCU) (see Z1 Truss Assembly), a power processing system used to coarsely regulate power from the arrays to 123 (+/– 2) volts DC. It has a maximum output of 6.25kW and is the source for all P6 secondary power.

The power storage system consists of a Battery Charge/Discharge Unit (BCDU) and two battery assemblies. They serve to charge the batteries during solar collection periods and to provide conditioned power to the primary power buses during eclipse. The BCDU measures 28in x 40in x 12in (71cm x 102cm x 30cm) and weighs 235lb (107kg). Each battery assembly consists of 38 nickel-hydrogen (Ni/H$_2$) cells with two connected in series able to store a total of 8kW of electrical power. The batteries have a design life of 6.5 years and can exceed 38,000 charge/discharge cycles at 35 per cent depth of discharge. Each battery measures 41in x 37in by 19in (104cm x 94cm x 48cm) and weighs 372lb (169kg).

The IEA electronics are kept at a safe operating temperature by a thermal control system comprising ammonia coolant, eight cold plates, two pump flow control systems, and one Photovoltaic Radiator (PVR). The cold plates are integral to the structural framework of the IEA with heat transferred to them via fine interweaving fins located on both the cold plates and the electronic boxes. The PV thermal control system is designed to dissipate on average 6kW of heat on each orbit and is commanded automatically by the IEA computer.

The PVR is deployed in orbit and consists of two separate flow paths through seven panels, each path independent. The PVR can reject up to 14kW of heat. It weighs 1,633lb (741kg) and when deployed from its concertina-like stowed position the seven hinged panels measure 44ft x 12ft x 7ft (13.4m x 3.7m x 2.1m). The entire P6 Truss and Solar Array Assembly has a mass on station of 34,866lb (15, 815kg).

STS-98/ISS 5A Destiny module

Atlantis
7 February 2001, 11.13am GMT
Cdr: Kenneth D. Cockrell (3)
Pilot: Mark L. Polansky
MS1: Robert L. Curbeam (1)
MS2: Marsha S. Ivins (4)
MS3: Thomas D. Jones (3)

The second Shuttle to visit the ISS during the tenure of Expedition 1, *Atlantis* lifted off with a total mass in the payload bay of 39,162lb (17,764kg), its primary objective being

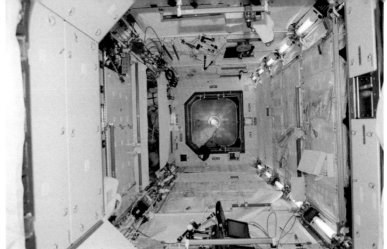

to carry the US Destiny science laboratory to the forward interface with Node 1, the Unity module attached to Zarya. Docking at the PMA-3 Node 1 nadir port occurred at an elapsed time of 1 day 17hr 38min, followed 19 hours later by a debris collision avoidance manoeuvre to slightly modify the orbit and avoid a potentially hazardous impact. Shortly thereafter, the Shuttle's remote manipulator arm was used to move PMA-2 from the forward port on Unity and re-site it to a temporary location on the Z1 truss on Unity's zenith (upper) docking position. While this was going on, Jones and Curbeam began the first of three EVAs by disconnecting cables and protective covers from Destiny prior to moving it to the ISS. The arm grappled Destiny 1hr 24min into the spacewalk and had it attached to the forward end of Unity 1hr 47min later. During the 7hr 34min spacewalk, Jones and Curbeam successfully hooked up cables and conduits linking Destiny to Unity.

Two days later the second spacewalk lasted 6hr 50min and saw Jones and Curbeam, assisted by the robot arm, move PMA-2 from the top of the Z1 truss to the forward end of Destiny, to which future Shuttle Orbiters would dock, after which they continued to make Destiny operational. Between spacewalks the crew were busy activating the new module and powering up systems and equipment connected internally and externally to Unity and the rest of the ISS, which by the addition of Destiny now exceeded Mir in the habitable volume throughout the full length of the facility. The third and last EVA, two days after the second, saw the same team attach a spare communications antenna to the outside of Destiny and release a radiator, inspecting the

solar arrays that proved troublesome to deploy on the preceding mission. They completed their tasks in 5hr 25min.

Only through this mission did the ISS come alive across all modules, with the CMG units on the Z1 truss now electrically active through Destiny. Mission 5A also delivered the capability to take over control of the station from the Russian Zvezda module and to interface with Russian systems to manage the environment, the thermal control systems, and attitude control. Across the duration of the mission the Shuttle conducted seven re-boost manoeuvres, leaving the ISS in a 212.5 x 199.2nm orbit. Apart from the Destiny module, *Atlantis* added 3,036lb (1,377kg) of dry cargo to the ISS and 368lb (167kg) via EVA, taking back 872lb (396kg) for return to Earth. Attached to the ISS for 6 days 21hr 15min, *Atlantis* undocked and got ready to return to Earth. But two extra days were taken up waiting for good weather before

ABOVE LEFT Destiny is lifted free of the Shuttle by Canada's remote manipulator arm and attached to the front of the Unity module. *(NASA)*

ABOVE With a uniquely 'clean' appearance, Destiny as it appeared when the Expedition 1 crew opened it up for use. *(NASA)*

BELOW Open for business! Destiny is formally accepted on orbit by the Expedition 1 crew. *(NASA)*

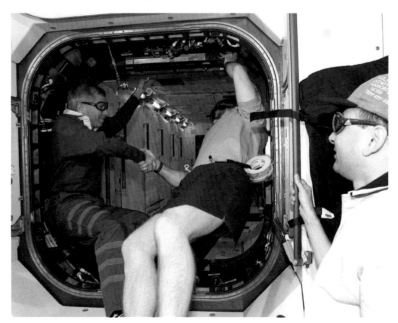

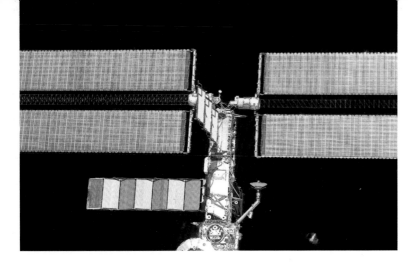

ABOVE Electrical power supplemented by the P6 array was in great demand as the physical volume of the ISS began to grow, a view seen during the visit of *Atlantis* on STS-98. *(NASA)*

BELOW The first US laboratory module, Destiny is prepared for launch aboard *Atlantis* in February 2001. *(Boeing)*

finally landing at Edwards Air Force Base in California due to inclement conditions in Florida, the mission having lasted 12 days 21hr 20min.

ANATOMY

Destiny module

Length	28.8ft/8.8m (30.2ft/9.2m with CBMs)
Diameter	14.6ft (4.5m)
Pressurised volume	3,750ft^3 (106.2m^3)
Weight	29,866lb (13,547kg) at launch and 52,962lb (24,023kg) fully outfitted
Launched	7 February 2001
Launch Vehicle	Shuttle *Atlantis* STS-98
Launch Site	Kennedy Space Center

The prime US research module, Destiny evolved from a much longer structure designed and planned as one of four zero-g facilities where US and international crews could conduct scientific and technological research. Those four long modules got cut to two in the drastic resizing brought about by a funding crisis in the 1990s. The two long modules were cut back in size to a single short module named Destiny with the Habitation Module deleted as a further cost-saving measure. Destiny became the first US laboratory in orbit since Skylab was occupied in 1974 – few would have thought it would be 27 years before NASA opened up its successor in space.

Construction began at the Marshall Space Center in 1995 shortly after the Alpha Station became the ISS and Destiny was delivered to the Kennedy Space Center three years later. Built by Boeing, Destiny consists of three cylindrical sections and two end cones, each of which contains a CBM hatch through which the astronauts can pass. Constructed of aluminium, the exterior uses a waffle pattern for strength and supports an insulation blanket of Nextel and Kevlar to alleviate the extremes of temperature experienced during day/night cycles each orbit.

Constructed of material similar to that used for bulletproof vests, the hull has an intermediate debris shield to protect the interior from micrometeoroids and space debris. This is surmounted by an aluminium

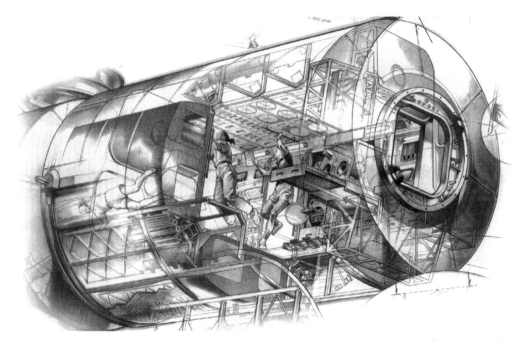

LEFT Considerably reduced in size when much of its systems were distributed to the Nodes, Destiny would be the core US element at the ISS. *(NASA)*

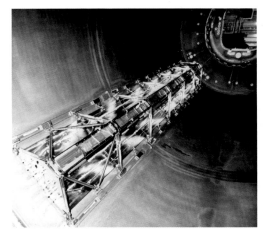

LEFT One of four stand-off structures in Destiny carrying conduits, piping, tubes and electrical cables to connect the module to other elements of the station. *(Boeing)*

BELOW Its cylindrical interior reshaped into a four-wall configuration by shaped experiment racks, Destiny is a place of scientific and technology research. *(NASA)*

debris shield for back-up protection and to reflect intense sunlight and relieve the load on the air conditioning system. The interior supports four stand-off structures, essentially long carry-through conduit trays for electrical power lines, data management cables, vacuum systems, air conditioning ducts, fluid lines, and interconnections for the station's experiment racks. A single 20in (51cm) diameter window is located on one side of the module. It is manufactured from optically clear glass and located in an open rack bay. It consists of four layers across a thickness of 5.6in (14.2cm) and has a removable outer shutter.

The interior of the module is rectangular and divided into four zones known as rack faces, essentially left, right, up, and down on all four flat longitudinal surfaces of the rectangular module. Each face can accommodate six racks of scientific equipment or experiments, making a total of 24 in all. Of these, 13 racks are removable and dedicated to specific scientific applications and 11 are equipped for power, cooling, temperature and humidity control, and for air revitalisation equipment to remove carbon dioxide and to replenish the module with oxygen. Eight rack bays are equipped with flexible curtains to provide 288ft³ (8.16m³) of temporary stowage space.

Destiny was equipped with only five systems racks when launched, including two avionics racks, two thermal control system racks, and an atmosphere revitalisation rack. The avionics

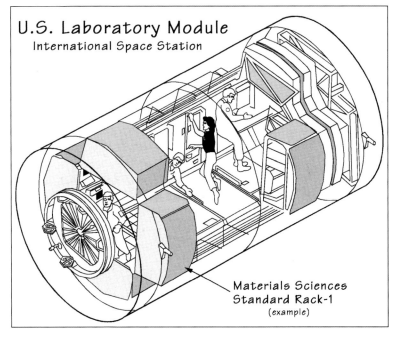

U.S. Laboratory Module
International Space Station

Materials Sciences
Standard Rack-1
(example)

RIGHT James Voss consults an atlas to refresh his knowledge on Asia and the Indian Ocean region as he makes 16 orbits of the Earth each day. *(NASA)*

racks help control the communications and tracking, environmental control system, thermal control, command data handling, and electrical power system. The thermal racks circulate chilled water to cool other racks and cabin air. One is a low-temperature system chilled to 4°C (39°F), providing cooling to selected racks – the other chills water to 17°C (63°F) for other racks and cabin air. The atmosphere revitalisation rack controls carbon dioxide removal and air monitoring. Later, logistics flights would bring up an additional four systems racks, additional science racks being outfitted as necessary to gradually expand the capabilities of the laboratory.

Eventually, Destiny would support the Minus Eighty Degree Laboratory Freezer (MELFI)

BELOW Expedition 2 crewmembers Yury Usachev, James S. Voss and Susan Helms were delivered to the ISS by *Discovery* in March 2001. *(NASA)*

lifted to the ISS by the STS-121 mission. After launch Destiny was attached to the forward docking interface of the Node 1 (Unity) module at one end, utilising a PCBM which berthed to Unity's ACBM. Eventually, Node 2 (Harmony) would be docked to the other end which carried an ACBM, Harmony itself having a PCBM. Eventually, Harmony would support the European Columbus and Japanese JEM modules, one either side with PMA-2 on the front to receive the Shuttle.

STS-102/ISS 5A.1 MPLM Leonardo/Expedition 2

Discovery
8 March 2001, 11.42am GMT
Cdr: James D. Wetherbee (4)
Pilot: James M. Kelly
MS1: Andrew S. W. Thomas (2)
MS 2: Paul Richards
MS3 (up): James S. Voss (4)
MS4 (up): Susan Helms (4)
MS5 (up): Yury Usachev (Russia)
MS3 (down): Sergei Krikalev (2) (Russia)
MS4 (down): William B. Shepard (3)
MS5 (down): Yuri Gidzenko (Russia)

The flight of *Discovery* was both a heavy logistics mission and an Expedition change-out flight, replacing Shepard, Gidzenko and Krikalev with Expedition 2 crewmembers Usachev, Voss and Helms. This was also the first Shuttle to carry the European Multi-Purpose Logistics Module (MPLM), a large cargo pod built by Italy and named Leonardo after the renaissance inventor, artist and sculptor Leonardo da Vinci. *Discovery* docked to the PMA-2 port on the front of the Destiny module at 1 day 19hr 16min, an early task being the relocation of PMA-3 from the Node 1 nadir berth around 90 degrees to the berth on the port (left) side, clearing that location for the MPLM.

Before that, Voss and Helms conducted an 8hr 56min EVA – the longest on record – to prepare PMA-3 for its relocation and to install a laboratory cradle from an ICC in the payload bay and a cable tray on the side of Destiny for later use. The EVA crew remained on standby in the Shuttle's airlock as PMA-3 was relocated. The MPLM was grappled by the robot arm at an

elapsed time of 3 days 15hr 54min and installed on Unity's nadir berth 2hr 50min later. A second EVA was conducted two days later by Thomas and Richards to install an External Stowage Platform (ESP) for spare parts and an ammonia coolant pump, completing the attachment of various cables and conduits set up on the first EVA. It lasted a more modest 6hr 21min.

A total of 9,649lb (4,377kg) was transferred to the ISS in addition to 980lb (445kg) of water in ten transfer bags, and *Discovery* returned with 1,647lb (747kg) of cargo and rubbish from the ISS. Several re-boost manoeuvres left the ISS in a 204.5 by 213.7nm orbit and the MPLM was re-berthed back in *Discovery*'s payload bay by 10 days 5min into the mission. The Shuttle had been docked to PMA-2 for 8 days 21hr 30min

when it undocked and returned to Earth at a mission elapsed time of 12 days 19hr 49min.

Expedition 2 was visited by Progress M1-6 and played host to three Shuttle missions – the last of which would return them to Earth – and to a unique flight designated ISS EP-1, which carried the first space tourist for a brief visit to the station. It supported three spacewalks – on 15, 18, and 21 July – logging 18hr 40min outside the ISS in completing installation of the Quest airlock, redeploying cables, and wiring harnesses. The last Expedition 2 EVA was the first to use the Quest airlock, during which nitrogen tanks were attached to the exterior. Expedition 2 ended after an ISS time of 163 days 8hr 13min when *Discovery* returned the crew to Earth at an elapsed time of 167 days, 6 hours, and 41 minutes.

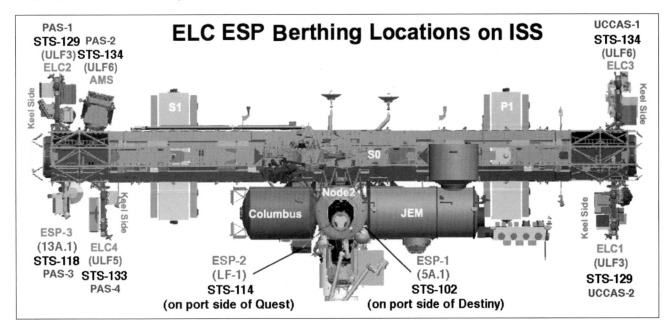

ELC ESP Berthing Locations on ISS

PAS-1
STS-129 PAS-2
(ULF3) STS-134
ELC2 (ULF6)
AMS

UCCAS-1
STS-134
(ULF6)
ELC3

Keel Side

S1

P1

Keel Side

S0

Keel Side

Node2

Columbus

JEM

Keel Side

ESP-3
(13A.1) ELC4
STS-118 (ULF5)
PAS-3 STS-133
PAS-4

ESP-2
(LF-1)
STS-114
(on port side of Quest)

ESP-1
(5A.1)
STS-102
(on port side of Destiny)

ELC1
(ULF3)
STS-129
UCCAS-2

Multi-Purpose Logistics Module (MPLM)

Length	21ft (6.4m)
Diameter	15ft (4.57m)
Weight	9,000lb (4,082kg)
Cargo capability	20,000lb (9,072kg)
Launch date	8 March 2001
Launch Vehicle	Shuttle *Discovery*

Three MPLM were built for the ISS by Alenia Spazio in Turin under contract to the Italian Space Agency (ASI) for ESA. Technically viewed as a descendant of Spacelab, which had been built for integral use with the Shuttle as a research laboratory mounted in the payload bay, the MPLM has the advantage of more recent engineering practices. New welding techniques have helped improve the structure/payload ratio in favour of a lighter shell with high cargo potential. More than 65 per cent of the fully loaded weight is in cargo and this adds great potential to the effectiveness of the MPLM for supplying the station with cargo. The big advantage was that the MPLM would be lifted from the Shuttle's cargo bay and temporarily mated to a Common Berthing Mechanism on the ISS, allowing stores and cargo to be floated across into the relevant modules.

Designed for a service life of at least 25 missions, the MPLM has provided Europe with a strong base of experience in supporting human space flight operations. MPLM shares much of its design and systems technology with the ESA Columbus laboratory module and this has helped keep costs down and improve its efficiency. It can support up to 16 International Standard Payload Racks (ISPRs) in its cavernous interior and when docked to the station the MPLM provides up to two weeks of storage and work volume.

For this mission the MPLM carried six systems racks for Destiny, two of which would contain power distribution equipment for laboratory payloads. Two of the racks supported a robotic work station required for controlling the ISS's robot arm, which was delivered on flight 6A. One rack carried equipment to activate the Ku-band communication system plus avionics to support the computer system. The final rack carried crew health equipment including a defibrillator and a respirator support pack.

The MPLM also delivered the first Human Research Facility (HRF-1), which would comprise the first experiments conducted on board. These would begin after the Shuttle departed. In addition the MPLM carried seven Resupply Stowage Racks (RSR) and Resupply Stowage Platforms (RSP), equipment necessary for crew rotation, spares for the station, food, and personal items. The racks remained attached to the MPLM but the items were collected up into transfer bags and transferred to the MPLM. Although built by the Italians the MPLM were owned and operated by NASA in exchange for research time aboard the ISS. The second was named Raffaello and the third Donatello.

External Stowage Platform 1 (ESP-1)

One of those components that are undersized for an oversized job, the ESP-1 installed on Destiny's port trunnion pin was the first of three external pallets carried to the ISS. Based on the design of the Integrated Cargo Carrier, the ESP is an unpressurised platform 8ft (2.44m) long by 1ft 6in (46cm) wide to which can be attached a range of Orbital Replacement Units (ORUs) and external subsystems necessary for the working operations of selected modules. ESP-1 carried a pump flow control system and a direct-

BELOW Yuri Gidzenko floats motionless inside the spacious Leonardo cargo module. *(ESA)*

current switching unit, powered by Unity. ORUs are attached to the ESP by Flight Releasable Attachment Mechanisms that connect them to a power supply tapped from the module to which the ESP is docked. ESP-1 was the smallest of the three carriers and was carried to the ISS attached to an Integrated Cargo Carrier, removed and located on Destiny during the second spacewalk.

STS-100/ISS 6A/ Canadarm2

Endeavour
April 19 2001, 6.41pm GMT
Cdr: Kent V. Rominger (4)
Pilot: Jeffrey S. Ashby (1)
MS1: Chris A. Hadfield (1) (Canada)
MS2: John L. Phillips
MS3: Scott E. Parazynski (3)
MS4: Umberto Guidoni (1) (Italy)
MS5: Yuri V. Lonchakov (Russia)

Endeavour conducted the ninth Shuttle flight to the ISS and the first to visit the Expedition 2 crew. It delivered the station's robotic arm that would be essential to all future assembly flights, a development of the Shuttle's Remote Manipulator System (RMS). Like the RMS had been for the Shuttle, the station's arm – also known as Canadarm2 – was the Canadian contribution to the ISS, the first of three robotic elements that would add greatly to the assembly and maintenance of the station. The second of three MPLM modules, this one – named Raffaello after Raffaello Sanzio, the great 15th-century Italian painter and architect – was carried to the station by Endeavour with cargo, water, and spares for the ISS. Raffaello had four Resupply Stowage Racks, four Resupply Stowage Platforms, and two scientific experiment racks (ExPRESS racks 1 and 2), one of which had a special vibration isolation system for payloads sensitive to motion. Total payload carried in the cargo bay was 38,330lb (17,386kg) with 6,346lb (2,878kg) of cargo and 1,380lb (626kg) of water transferred to the ISS; 1,608lb (729kg) was returned in Endeavour.

Hadfield and Parazynski conducted two spacewalks, the first of which took place on the day after Endeavour docked at PMA-2. As

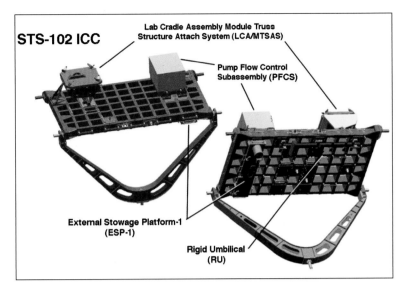

the two got ready for their EVA, the Shuttle's robotic arm lifted a Spacehab pallet containing the SSRMS and a UHF communications antenna from Endeavour and moored it to a fixed cradle on Destiny. Working from a foot platform at the end of the Shuttle's manipulator arm, Hadfield unpacked the restraints and cables that held the SSRMS in a folded configuration and connected power cables running up through the grapple fixture on Destiny to which it would be attached. This served as the initial operating base for the SSRMS, which would not be fully mobile and completely operational until the two additional elements of Canadarm2 had been delivered by later Shuttle missions.

The two astronauts then removed the UHF antenna and installed it on the station, releasing its boom before diverting their attentions back to Canadarm2. By releasing 32 jack bolts they gained access to eight 4ft (1.2m) long restraining bolts locking the arm to the pallet. When those had been removed and stowed on the pallet for return to Earth, Canadarm2's dextrous arms could be unfolded and fixed rigidly in position, ending a 7hr 10min spacewalk in which Hadfield became the first Canadian to perform an EVA. The following day the Shuttle crew joined the Expedition 2 astronauts aboard the ISS and the Raffaello logistics module was removed from the cargo bay and attached to the nadir port on the Unity module, where its logistical supplies could be offloaded. It would be returned to the Shuttle's cargo bay for return to Earth.

ABOVE The Integrated Cargo Carrier (ICC) was a useful bridge within the Shuttle's cargo bay for supporting packages and platforms for transfer to the station. (NASA)

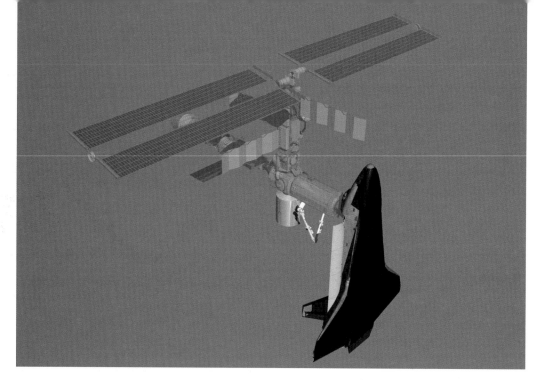

Connecting the power data grapple fixture circuits from Destiny to Canadarm2 during a second EVA lasting 7hr 39min, the two spacewalkers completed the SSRMS installation for this flight. Later in the mission, still attached to the pallet at one end, the other end of Canadarm2 grasped a fixture on the side of Destiny. Thus moored with one fixture, the 'free' end was used to grasp the Spacehab pallet, release it from its temporary location on the outside of Destiny, and pass it to the Shuttle's arm which then placed the rack back in the cargo bay, making this the first

robotic 'handover' in space! Two ISS re-boost manoeuvres were conducted by *Endeavour* before it undocked after 8 days 3hr 23min, returning back to Earth at an elapsed time of 11 days 21hr 30min.

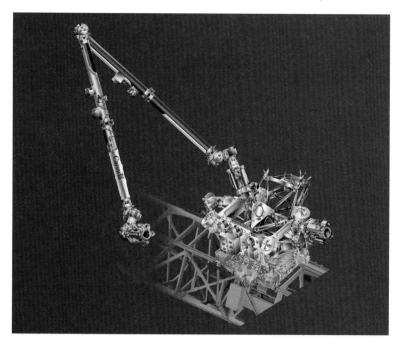

ANATOMY

Space Station Remote Manipulator System (SSRMS)

Designed and manufactured by MDA Space Missions of Canada, formerly Spar Aerospace, the SSRMS is affectionately known as Canadarm2, being the second space-qualified articulating arm designed by this company. Early in the development of the Shuttle, Canada was awarded responsibility for the Orbiter's Remote Manipulator System (RMS), and that proven and reliable arm has been utilised on the majority of Shuttle flights. For the station, the technical requirement was much greater and the mass of the loads it would handle were considerably greater. Moreover, the operating flexibility inherent in the design requirement was a generation beyond that achieved with the Shuttle's RMS.

The SSRMS was only the first of three robotic support elements lifted to the station for integrated operation, relieving station crewmembers of many hours of spacewalks and simplifying essential mechanical tasks outside the pressurised compartments of the ISS. The second element, the Mobile Remote Servicer Base System (MBS), was added on the

RIGHT The Station Manipulator System is able
to 'walk' across from one structure to another by
connecting to power pickup points. (CSA)

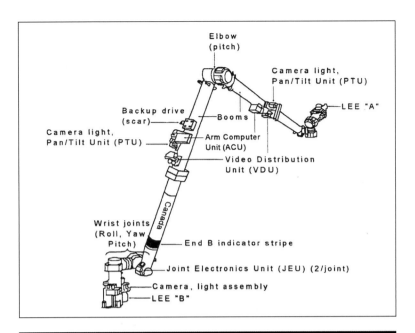

STS-111 mission just over a year later followed
by the Special Purpose Dextrous Manipulator
(SPDM, affectionately known as 'Dextre') during
STS-123 in March 2008. The MBS would
provide greater access by affording Canadarm2
the full range of the truss assembly and Dextre
would be attached to the other end for a
superior range of tasks.

In the interim, Canadarm2 attached itself
to a Power Data Grapple Fixture (PDGF), one
of several located at different places around
the station's exterior. These enabled it to inch
its way across the station and hook on to any
PDGF attached to the exterior surface of the
modules. Each PDGF has a central grapple pin

RIGHT Installed as Canadarm2, the manipulator
is an integral part of station operations and only
the first in a continuing series of robotic devices.
(NASA)

BELOW The hand, or 'end effector', of the
SSRMS carries a range of sensors and drive
systems together with an effective snare device
for latching to payloads or base points. (CSA)

for physically securing the arm to the station using a Latching End Effector (LEE) attached at both ends of the arm. The LEE is essentially an anchor point for the arm, allowing it to flip end-over-end, 'walking' across grapple fixtures to which it can attach itself. Four rectangular-shaped electrical connectors are inside each PDGF for transferring power and data to the arm or to a powered object on the extreme end of the arm.

Canadarm2 is controlled from either one of two Robotic Work Stations (RWS), one situated in Destiny and the other in the Cupola attached to the Tranquility (Node 3) module berthed to the nadir port on Unity (Node 1). Astronauts using conventional hand controllers and laptops can view the work area and control the motion of the manipulator. Support and technical back-up to Canadarm2 is provided by the Canadian Space Agency through its Space Operations Support Center on the ground, which has all the appearance of a mini-Mission Control. Canada also funds a Mission Operations and Training Simulator at Saint-Hubert for training astronauts in the basics of robotic operations. In return for providing the arm and facilities, Canada gets 2.3 per cent of designated research volume in non-Russian modules and is entitled to send one astronaut to the station every three years on a tour lasting three to four months.

A challenge in designing Canadarm2 resulted from its location in the Shuttle for delivery to orbit. Because it weighs more than four times as much as the Shuttle RMS, Canadarm2 had to be put together so that it would withstand

the physical loads of launch and ascent. Unlike the Shuttle RMS, Canadarm2 had to be folded into a U-shape on a Spacehab pallet across the width of the Shuttle's cargo bay, imposing transverse loads not experienced by the Shuttle's arm, which is supported along its full length down one side of the Orbiter's cargo bay. Moreover, a fundamental planning decision in streamlining and simplifying the way the ISS is operated provided numerous packages attached to the outside of the station to replace failed units as necessary. These are known as Orbital Replacement Units (ORUs), self-contained equipment that Canadarm2 can swap as subsystems wear out or fail. And because Canadarm2 will never return to Earth, there are ORUs for the arm itself to keep it functional for several decades.

Because Canadarm2 is not permanently attached to one fixture, unlike the arm used by the Shuttle it can move around outside at the will of the astronauts, or when attached to the MBS it can travel the entire length of the truss assemblies. Like a human arm it has shoulder, elbow, and wrist articulation, but unlike humans it has three joints at the shoulder position, one at the elbow, and three at the wrist. This compares to just two shoulder joints on the Shuttle arm, one at the elbow, and three at the wrist. The Shuttle arm has six degrees of freedom but Canadarm2 has seven, and whereas the RMS has elbow rotation limited to 160 degrees, Canadarm2 has 540 degrees of rotation on all seven joints, providing much greater freedom than a human arm. The Shuttle RMS has no sense of touch but Canadarm2 has feelers to measure the intensity of contact, with an automatic vision feature for capturing free-flying payloads and a collision avoidance capability to prevent a high rate of contact damaging the arm or the object encountered.

With a length of 57ft 9in (17.6m), compared to the Shuttle arm's length of 50ft 3in (15.3m), it is fabricated from 19-ply, high-strength carbon-fibre thermoplastic. Canadarm2 has greater reach, with an external diameter of 14in (36cm), only 1in (2.5cm) greater than that of the RMS, and has four colour cameras, one each side of the elbow and one at each LEE, compared to one each on the elbow and the wrist of the Shuttle arm. One of the greatest challenges to

its stability and rigidity was the greater mass Canadarm2 was capable of moving, some 255,700lb (115,984kg) compared to 66,000lb (29,937kg) by the Shuttle arm. Although weightless in orbit, the mass has inertia and therefore has to handle that equivalent load.

Unloaded, Canadarm2 can move at 15in (38cm) per second compared to a maximum motion of 2ft (61cm) per second for the RMS. Loaded, the respective figures are less than 1in/sec compared to 2in/sec. During spacewalks, Canadarm2 can move at 6in/sec or less than 0.5in/sec with a 200,000lb (90,718kg) mass on the end. All this size and complexity has a price, however – Canadarm2 weighs 3,960lb (1,796kg) compared to 905lb (410kg) for the Shuttle arm but it is comparatively efficient, taking an average 435W to keep it alive or 2kW at peak operation.

Soyuz TM-32/ISS 2S (EP-1)

28 April 2001, 7.37am GMT
Cdr: Talgat Musabayev
Flight Engineer: Yuri Baturin
Tourist: Dennis Tito (US)

Just a few hours after *Endeavour* undocked from the ISS, the world's first orbital tourist docked to Zarya's nadir (Earth-pointing) port, the American millionaire Dennis Tito, who financed his own trip. Initially arranged with the Russian Space Agency through MirCorp, Tito was originally set to train in the US to be taken to the Mir station. But when the three crewmembers arrived in America for check-ups and rudimentary familiarisation they were sent back, with NASA manager and ex-astronaut Robert D. Cabana stating they were 'not willing to train Dennis Tito'.

NASA has always opposed fee-paying tourists, fearing that a lack of professional experience and years of training could make them a danger. Moreover, people aboard the ISS consume atmospheric gases, food, and water, and have to be chargeable to what is in essence an international government research laboratory. A heated stand-off delayed the venture until Space Adventures Ltd renegotiated the deal and Tito was signed to fly to the ISS instead of Mir when the latter was abandoned to a fiery re-entry.

The three-man crew of Soyuz TM-32 docked with the ISS at the Zarya nadir port at 7.58am GMT on 30 April, joining the Expedition crew inside the ISS. For several days the six people aboard the ISS conducted their own activities and Tito got his space flight, undocking and returning to Earth in Soyuz TM-31 after a mission lasting 7 days 22hr 4min, leaving their spacecraft as the return ISS lifeboat. The experience cost Tito $20 million. This flight is generally regarded just as Soyuz TM-32 but it is also referred to as EP-1, the first in a succession of tourist visits to the station.

STS-104/ISS 7A/Quest joint airlock

Atlantis
12 July 2001, 9.04am GMT
Cdr: Steven W. Lindsey (2)
Pilot: Charles O. Hobaugh
MS1: Michael L. Gernhardt (3)
MS2: Janet L. Kavandi (2)
MS3: James F. Reilly (1)

Atlantis made the tenth Shuttle visit to the ISS, providing the station with the Quest airlock capable of supporting spacewalks directly from the ISS itself using Russian Orlan or US space suits and incorporating a new pre-breathing protocol. It was also the first mission supported from the new ISS Flight Control Room at Houston's Johnson Space Center. Originally planned for mid-June, the flight had been put on hold when problems were encountered with Canadarm2 aboard the ISS. The arm would be the only means of lifting Quest into position so its satisfactory operation was crucial to the mission. The glitch was traced to a back-up mechanism in the arm's shoulder pitch joint mechanism.

The ISS had already grown into a comprehensive station when the Shuttle arrived two days after launch. One Progress vehicle was docked to the aft end of Zvezda with a Soyuz lifeboat attached to Zarya's nadir port, with *Atlantis* being the third spacecraft at the station attached to the PMA-2 port on the front of Destiny. Quest was moved from the Shuttle's cargo bay and transferred to the starboard (right) port on Unity, directly opposite PMA-3,

ABOVE **Canadarm2 manoeuvres the Quest airlock from *Atlantis*'s cargo bay during the delivery flight, July 2001.** *(NASA)*

BELOW **Quest attached to the starboard docking port on Node 1, an integral and permanent part of the station.** *(NASA)*

on the first EVA, which began at an elapsed time of 2days 18hr 7min.

Operated from the aft flight deck by Kavandi, the Shuttle's arm was used to move Reilly across to Quest, still latched down in the cargo bay, where he removed covers and installed fixtures that would be used on the second and third EVAs to wire the airlock to its permanent position on the ISS. But the Shuttle's arm did not have sufficient reach to move Quest off the cargo bay and on to its place on the ISS, so from the Destiny module Expedition 2 science officer Helms directed Canadarm2 to

do that job, grasping it with the arm 1hr 39min into the EVA. Lifting Quest from the Shuttle, it was slowly moved across and attached to Node 1 on the starboard docking port 2hr 56min later. Fellow spacewalker Gernhardt accompanied Reilly to inspect the docking mechanism on Unity before it was locked down with 18 motorised bolts, wires, and cables, then installed to power Quest's heating system.

The EVA lasted 5hr 59min but when activation of Quest began inside Unity problems arose with a suspected pressure drop in the airlock's coolant system, misinterpreted by the software as a water leak. Fixing that took time and delayed the second spacewalk but the mission was extended by a full day to provide more time. The second EVA lasted 6hr 29min and had Gernhardt stationed on the Shuttle robotic arm supporting the movement of one set of oxygen and nitrogen tanks from the Spacehab pallet, a task performed by Helms working Canadarm2 from the space station, with Reilly positioned atop Quest on the Unity node. Both astronauts used foot platforms on respective arms to install and align the tanks, attaching hoses and cables from the tanks to the airlock.

All EVAs for ISS assembly had taken place through the Shuttle's airlock module, including the first two on this mission, but the third EVA would be a trial run of the new Quest airlock attached to Unity. In a 4hr 2min spacewalk, Gernhardt and Reilly conducted the first EVA from the new joint airlock and supervised the move of Quest's second set of oxygen/nitrogen tanks from the Shuttle's cargo bay across to the top of the module, fully operational for depressurising and repressurising the airlock for future spacewalks from the ISS. After 8 days 1hr 49min docked to the station, *Atlantis* moved away and returned to Earth, landing at an elapsed time of 12 days 18hr 35min, leaving the Expedition 2 crew to prepare for handover to the Expedition 3 team just a few weeks later.

ANATOMY

Quest joint airlock

A fundamental requirement of servicing and maintaining the ISS is that astronauts should be able to conduct spacewalks and gain access to the exterior. Any dedicated ISS airlock must also be able to support the different

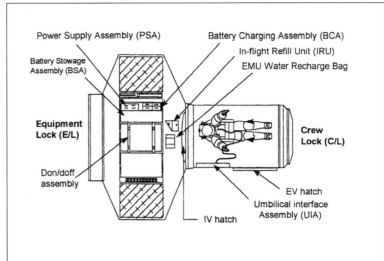

Power Supply Assembly (PSA) Battery Charging Assembly (BCA)

Battery Stowage Assembly (BSA) In-flight Refill Unit (IRU)

EMU Water Recharge Bag

Equipment Lock (E/L)

Crew Lock (C/L)

Don/doff assembly

EV hatch

Umbilical interface Assembly (UIA)

IV hatch

requirements of the Russian Orlan suit and the US Extravehicular Mobility Unit (EMU). Russian airlock hatches are too small for NASA astronaut suits and have different fixtures and fittings to support the preparations for EVA. It was essential to develop an airlock both could use and to have that available as an early addition to the station. Until the Quest airlock was installed by STS-104 NASA astronauts had to use the Shuttle docking module and associated airlock, from which the purpose-built Quest design was derived, while Russian EVAs could only be conducted by them through the Transfer Chamber on Zvezda.

Quest is a simple design comprising a large diameter Equipment Lock (EL) and a somewhat narrower Crew Lock (CL). Together they provide the means to store and don Russian Orlan and US EMU suits and to provide space and facilities where they can be maintained. The Equipment Airlock has stations that allow the astronauts and cosmonauts to don and doff their suits and it is equipped with two racks. One rack is for avionics and the other for cabin air, with batteries, power tools, and other EVA supplies housed inside. The Crew Lock incorporates an Umbilical Interface Assembly which provides communication equipment and space suit power interface for two astronauts simultaneously.

Quest also has external handrails, lighting, and mountings for four 3ft (91cm) diameter carbon-fibre tanks, each with a volume of 15.1ft^3 (0.43m^3) and with a weight of 1,200lb (544kg). They comprise two nitrogen and two oxygen tanks for atmospheric replenishment

and as part of the High Pressure Gas assembly. A fundamental function of the Quest airlock is to provide a 'camp-out' capability to reduce the time astronauts need to pre-breathe oxygen to prevent attacks of the bends. The basic difficulty is that with a mixed oxygen/nitrogen atmosphere at sea-level pressure, a degree of pre-breathing on pure oxygen is necessary to prevent nitrogen bubbles gathering inside the body.

The Russian protocol is to breathe pure oxygen for 30 minutes in the suit but the NASA method adopted for Shuttle operation was more complex, calling for a reduction in cabin pressure from 14.7lb/in^2 to 10.2lb/in^2 for at least 36 hours prior to EVA, breathing 100 per cent oxygen in the suit for 40 minutes. Because of its greater volume the ISS must maintain a 14.7lb/in^2 atmosphere throughout, necessitating a different protocol.

Camp-out begins the day prior to the scheduled EVA, when crewmembers breathe 100 per cent oxygen before their sleep period, when they are locked in the Quest Equipment

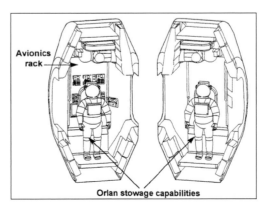

Avionics rack

Orlan stowage capabilities

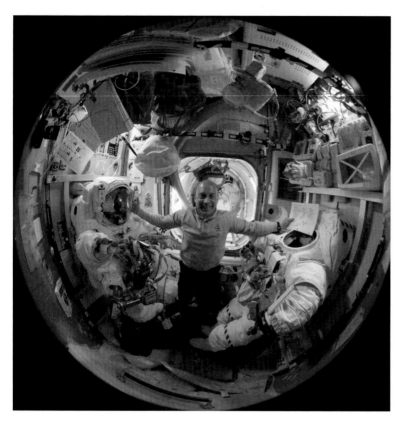

Endeavour
10 August 2001, 9.10pm GMT
Cdr: Scott J. Horowitz (3)
Pilot: Frederick W. Sturckow (1)
MS1: Patrick G. Forrester
MS2: Daniel T. Barry (2)
MS3 (up): Frank L. Culbertson Jr (2)
MS4 (up): Vladimir N. Dezhurov (1) (Russia)
MS5 (up): Mikhail Tyurin (Russia)
MS3 (down): James S. Voss (4)
MS4 (down): Susan J. Helms (4)
MS5 (down): Yuri Usachev (1) (Russia)

ABOVE A fish eye lens view of the interior of Quest with Garrett Reisman displaying its relatively spacious interior. *(NASA)*

Lock which is depressurised down to 10.2lb/in^2. After an eight-hour sleep the astronauts repressurise Quest back up to 14.7lb/in^2 with the crew on oxygen masks for an additional hour during their hygiene break aboard the ISS. After that they are sealed up again and don their suits, thus only needing to have a 30 minute in-suit pre-breathe just prior to depressurising Quest, first down to 3lb/in^2 to check possible suit leaks, and then opening up the Crew Lock and going outside. ISS crewmembers operate their suits at an internal pressure of 4.3lb/in^2; anything higher and the suits stiffen and quickly tire the occupant.

Even with the advantage of the new Quest airlock, the amount of time spent getting ready for an EVA far exceeds the time spent productively outside and the impact on station operations and scientific activity is great. This is the reason why the Canadarm2 robotic manipulator can replace the human function for so many external tasks and why robotics in general is a great asset on space station operations.

Quest is 18.5ft (5.64m) long with a diameter of 13ft (4m) across the Equipment Lock and with an internal volume of 1,200ft^3 (33.98m^3). It is fabricated from aluminium and weighs 13,368lb (6,064kg).

This was another resupply and Expedition crew rotation flight with *Endeavour* going back to collect Expedition 2 and to leave Expedition 3 astronauts for their 125-day stay aboard the station. *Endeavour* carried the Leonardo MPLM on its second flight stocked with six Resupply Stowage Racks, four Resupply Stowage Platforms, and two science racks for Destiny – ExPRESS racks 4 (1,180lb/535kg) and 5 (1,200lb/544kg). The total weight of racks and associated equipment uploaded to the station was 11,000lb (4,990kg) of which 6,770lb (3,070kg) was cargo. In all, 10,651lb (4,831kg) was transferred to the ISS, 9,657lb (4,380kg) of which was logistics, with 3,802lb (1,725kg) transferred back to the Shuttle. Of this, 2,564lb (1,163kg) was in the Leonardo module.

Another payload was the Materials International Space Station Experiments (MISSE) package for external mounting to test and evaluate the effects of the space environment on a wide variety of different materials. These were packaged in four Passive Experiment Containers and had been used initially for experiments on Mir during 1996. As with all missions to the ISS, the Shuttle carried a wide range of small experiments and research equipment. Two spacewalks were conducted by Sturckow and Forrester, lasting 6hr 16min and 5hr 29min respectively, during which the MISSE packages were attached to the exterior of the Quest airlock and additional rails and handholds were installed for later visits. After a docked duration of 7 days 19hr 48min, *Endeavour* returned to land 11 days 21hr 12min after launch.

Before they returned to Earth aboard *Endeavour* in December 2001, completing their mission of 124 days, 22hours and 47minutes, the Expedition 3 crew would receive a Soyuz visit in October and two Progress cargo-tanker flights in addition to the delivery of Russia's Pirs module. In addition, they conducted four spacewalks on 8 October, 15 October, 12 November, and 3 December, totalling 17hr 50min, mostly concerned with maintenance tasks on Pirs (launched 14 September) and Zvezda, as well as attending to obstructions on a Progress docking port. Their space flight duration of 128 days 20hr 46min ended on 17 December when they landed aboard *Endeavour*.

The most memorable event for NASA astronaut Frank Culbertson was when he saw from space the appalling aftermath of the attack on the Twin Towers in New York on 11 September. Grabbing a camera, he took pictures of the dense pall of smoke drifting away from the city as word came up from Mission Control, keeping both he and his Russian crewmates aware of developments. Culbertson would use his experience as the only American not on Earth when this tragic event happened to appeal for unity and tolerance among all people.

Pirs/ISS 4R

14 September 2001, 11.35pm GMT

Following the launch of Zarya in 1998 and Zvezda in 2000, Pirs was Russia's third module delivered to the ISS, launched by a Soyuz rocket. Developed by RKK Energia, Pirs is also known as Docking Compartment 1 (DC-1), providing a third port for Soyuz and Progress vehicles in addition to the aft port on Zvezda and the nadir port on Zarya. The only other two docking ports on the Russian components were the zenith and nadir ports on Zvezda, both designed for larger and heavier modules. Previously, whenever a new Soyuz arrived at the ISS the unmanned Progress cargo-tanker had to be separated and flown in formation with the station until the preceding Soyuz had returned its crew to Earth, vacating a port for the cargo-tanker to re-dock.

Pirs incorporated an airlock for two Orlan-

LEFT Russia's **Pirs module was delivered to the ISS in September 2001, providing an additional docking walk and airlock facility for Russian space walks.** *(NASA)*

suited cosmonauts to conduct spacewalks from the Russian components of the station. The Pirs module rendezvoused with the ISS on 16 September 2001 and docked to the Zvezda nadir port at 1.05am GMT on 17 September. Additional docking modules were sent up in 2009 and 2010 to increase that to four and to add an extra airlock.

ANATOMY

Pirs

Length	16.1ft (4.91m)
Diameter	8.4ft (2.56m)
Weight at launch	9,590lb (4,350kg)
Weight in orbit	7,893lb (3,580kg)
Payload capacity for deliverable cargo	1,764lb (800kg)
Pressurised volume	459ft³ (13m³)

The Docking Compartment (DC-1) was first designed to serve the Mir-2 station but with the merger of Alpha One and the Russian successor to Mir, DC-1 became an integral part of the Russian side of the ISS. It was designed in the Russian tradition of giving the unmanned module autonomous manoeuvring capabilities via an instrument and propulsion section adopted directly from an existing spacecraft. Designated PAO, this section of the module was essentially a Progress equipment and propulsion section and carried Pirs to the station. Jettisoned after it arrived at the ISS, the PAO exposed the docking hatch that would be used to receive future Soyuz craft.

DC-1 serves as both docking module and airlock assembly, with two hatches for two spacewalkers to egress and ingress without depressurising the ISS. In that function it is similar to the US Quest airlock. It can also move propellant from the tanks of a docked Progress vehicle to tanks in either Zvezda or Zarya and can, in reverse flow, provide propellants from those modules to docked Soyuz and Progress vehicles through hard-line connections and electrical feed lines.

Pirs was designed for a five-year lifetime and was scheduled to have been replaced by Russia's Multipurpose Laboratory Module, which would have had integral airlock and docking facilities of its own. Financial difficulties brought cancellation of that plan with an

BELOW Pirs was attached to the Zvezda nadir (Earth facing) port in September 2001, the first of three new such facilities that would eventually be delivered to the ISS. *(NASA)*

additional docking facility provided by Poisk (MRM-2), launched to the Zvezda zenith position on 10 November 2009, and by Rassvet (MRM-1), which was carried aboard *Atlantis* on 14 May 2010.

Soyuz TM-33

21 October 2001, 8.59am GMT
Cdr: Viktor Afanaseyev (Russia)
Flight Engineer: Claude Haignere (France)
Engineer: Konstantin Kozeyev (Russia)

This ISS taxi flight carried cosmonauts Afanaseyev and Kozeyev and French astronaut Haignere to the station, who returned with the TM-32 spacecraft left by the tourist flight in April. TM-33 docked to the Zarya nadir port at 10.00am GMT on 22 October and the three crewmembers joined Expedition 3 for more than seven days of joint activity. Relocating their contour couches to TM-32, the short-stay visitors undocked at 1.39am GMT on 31 October, landing at 4.58am GMT after a flight lasting 9 days 17hr 59min.

STS-108/ISS UF-1/ Expedition 4

Endeavour
5 December 2001, 10.19pm GMT

Cdr: Dominic L. Gorie (2)
Pilot: Mark E. Kelly
MS1: Linda M. Godwin (3)
MS2: Daniel M. Tani
MS3 (up): Carl E. Walz (3)
MS4 (up): Daniel W. Bursch (3)
MS5 (up): Yuri I. Onufrienko (Russia)
MS3 (down): Frank L. Culbertson Jr (2)
MS4 (down): Vladimir N. Dezhurov (1) (Russia)
MS5 (down): Mikhail Tyurin (Russia)

Endeavour returned to space to exchange long-stay crewmembers and to carry the MPLM Raffaello, which was packed with a wide range of experiments and logistics equipment, transferring 6,244lb (2,832kg) to the ISS and returning with 4,156lb (1,885kg). But STS-108 was a maintenance mission as well, and a single 4hr 11min EVA from the Shuttle airlock module allowed Godwin and Tani to place thermal insulation blankets around the two cylindrical mechanisms that rotate the solar arrays, conduct a few minor repairs, and stash tools for future spacewalks. *Endeavour* carried into space flags from the three sites hit by terrorists on 9/11, including the Stars and Stripes wrapped around a flagpole near the site of devastation in New York, plus a US Marine Corps flag from the pentagon and a flag from the State capital in Harrisburg, PA, along with 6,000 smaller flags that were distributed to families that lost loved ones that day. The mission lasted 11 days 19hr 37min.

The Expedition 4 crew of Yuri Onufrienko, Daniel Bursch, and Carl Walz formally began their mission on 7 December at 8.03pm GMT, and all three participated in the two-man EVAs, the first of which lasted 6hr 3min on 14 January 2002, during which the cargo boom for the Strela crane was relocated from PMA-1 to the Pirs module. A second EVA on 25 January lasted 5hr 59min, when jet efflux deflector shields were attached to the Zarya module. Unlike the first two spacewalks, conducted from the Pirs module, the third EVA on 20 February was conducted through the Quest airlock. It lasted 5hr 49min and

qualified the facility for four spacewalks scheduled for the STS-110 mission. Expedition 4 ended when they were relieved by the fifth long-stay crewmembers launched aboard *Endeavour* in June 2002. Their mission had lasted 190 days 5hr 31min, and they had been in space for 195 days, 11 hours and 38 minutes.

STS-110/ISS 8A/S0 Truss/ Mobile Transporter

Atlantis
April 8 2002, 8.44pm GMT

Cdr: Michael J. Bloomfield (2)
Pilot: Stephen N. Frick
MS1: Rex J. Walheim
MS2: Ellen Ochoa (3)
MS3: Lee E. Morin
MS4: Jerry L. Ross (6)
MS5: Steven L. Smith (3)

Atlantis delivered the S0 truss to the top of the Destiny module. Arguably the most important element for the successful long-term development of the ISS, this massive structure would form the central backbone to which additional truss sections would be attached, growing out from this centre segment to eventually span more than 350ft (107m) and

ABOVE The three crewmembers of Expeditions 3 were exchanged for the three members of Expedition 4 with the four-person crew of *Endeavour* in December 2001. *(NASA)*

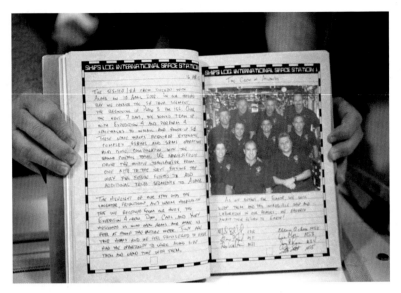

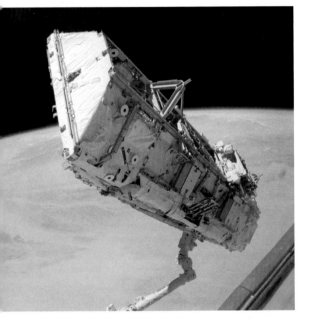

ABOVE LEFT Is a spacecraft a 'ship'? Some believe it is and if that is so, it needs a ship's log – which is exactly what this is recording the visit of *Atlantis* in April 2002, kept aboard the ISS. *(NASA)*

ABOVE RIGHT With Soyuz attached to Zarya's zenith port and a Progress tanker docked to Zvezda, the transverse S0 truss looms large when viewed through the solar panels on the Russian sector, the Z1 truss in foreground supporting the P6 array. *(NASA)*

LEFT Manoeuvred into place by the remote manipulator arm, the S0 truss assembly would form the permanent backbone to the station. *(NASA)*

BELOW LEFT Installation crewmembers work to attach bracing struts connecting the S0 truss to Destiny forming a rigid attachment capable of withstanding bending moments. *(NASA)*

BELOW RIGHT Work space around the S0 truss gets crowded with the adjacent P6 solar array and its delicate, ultra-thin solar cell panels. *(NASA)*

support all eight Solar Array Wings. Also flown up to the station was the Mobile Transporter (MT) to give Canadarm2 much needed mobility along the truss assembly. The crew selected for this flight were chosen for four long and arduous spacewalks and assembly activity. Between them, STS-110 spacewalkers Ross and Smith already had 12 EVAs under their belts, but the mission would also give Walheim and Morin their first EVAs and their first space flights. *Atlantis* docked to PMA-2 at 4.19pm on 10 April.

One of the first tasks was to use Canadarm2 to lift the S0 truss from the Shuttle and latch it to a Lab Cradle Assembly on top of Destiny, mooring it with sufficient rigidity through securing struts. These were attached during the first EVA, which began at an elapsed time of 2 days 17hr 52min and lasted 7hr 48min. Together, Smith and Walheim set up the securing struts and began harnessing the new segment to electrical connectors on Destiny, which was continued two days later when Ross and Morin spent 7hr 30min completing the installation of truss S0. Next day Smith and Walheim took a 6hr 27min walk to reconfigure Canadarm2 and began to remove launch restraints and protective covers from the Mobile Transporter, unfolding and installing the Airlock Spur. The fourth EVA took place two days after the third, and Ross and Morin spent 6hr 37min finishing up the attachment of cables and umbilicals, preparing for the next mission, and making the MT operational. In his two EVAs on this mission, Jerry Ross raised his total spacewalk time to 58hr 18min, a US record.

After being docked for 7 days 2hr 12min, *Atlantis* returned to Earth 10 days 19hr 43min after launch. Including the S0 truss, *Atlantis* had offloaded 30,600lb (13,880kg), including 146lb (66kg) of oxygen, 45lb (20kg) of nitrogen, and 1,465lb (665kg) of water, and received back 2,607lb (1,183kg) for return to Earth.

ANATOMY

S0 Central Integrated Truss Structure

Length	44ft (13.4m)
Width	15ft (4.6m)
Weight	27,830lb (12,623kg)

The S0 truss assembly was manufactured by Boeing at Huntington Beach, California, former home of McDonnell Douglas, which had done

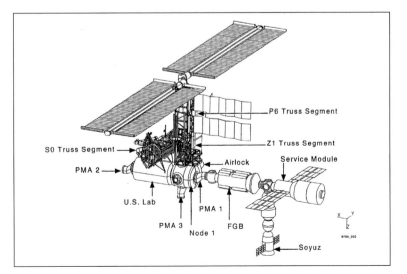

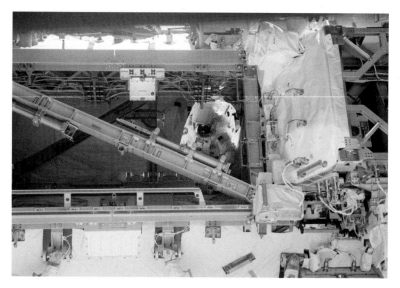

so much work on so many former space station concepts over the previous decades. S0 would provide the structural backbone for the power generation subsystem of the ISS electrical system and provide power conversion and routing capabilities to downstream loads. It would carry information to the station's extremities where the solar arrays would be located and it would provide power for experiments and a wide variety of systems and subsystems using the truss assemblies for location and operation.

The S0 structure consists of an elongated hexagonal cross-section with five bays arranged along the longitudinal axis. Most of the equipment had been built in to the truss during manufacture. The frame is open enough, however, to allow an astronaut to gain access

to the interior through spaces in the bays, their numbered sections making navigating to a specific location easy, which allows an astronaut to move around the truss. The faces are numbered sequentially counter-clockwise from the forward nadir (Earth-pointing) face. S0 was preassembled with all the equipment necessary to support four photovoltaic module truss segments to increase power production including a Mobile Transporter for the Canadarm2, a portable work platform, four GPS antennas, umbilicals for connecting the US modules, main bus switching units, circuit interrupter devices, and an airlock spur to link Quest with the S0 truss for spacewalkers.

As the centre segment of 11 integrated trusses that would be brought to the ISS, it acts as a junction from where external utilities are routed to the pressurised modules, including an active thermal control system using ammonia. The aft face of the S0 supports a 21ft (6.4m) radiator panel which transports heat to the S0 electronics boxes through a system of internal pipes. The forward face carries rails to form a track along which the Mobile Transporter can move up and down, additional track being laid with each new truss segment growing outwards from this centre section. Attachment of these additional truss segments would be made by a remotely controlled capture latch and four motorised bolt assemblies on each side. Structurally attached to the Destiny module by ten telescoping struts set up on the spacewalks conducted by STS-110 crewmembers, the S0 would remain permanently attached to the station.

The Module-to-Truss Segment Attachment System (MTSAS) includes the Lab Cradle Assembly (LCA), a soft-capture capability for mating the truss structure to Destiny. The MT struts for attaching S0 to Destiny were preassembled and supported the other ten segments installed on subsequent flights. The LCA consists of a capture and alignment mechanism and a rigid capture bar. During mating the capture claw engages the bar on the S0 passive shelf and draws the truss segment down on to the LCA interface for a semi-rigid structural hold. Four umbilical trays carry two fluid and two power and data lines for thermal protection, communications, and electrical interfaces with other elements. The port and

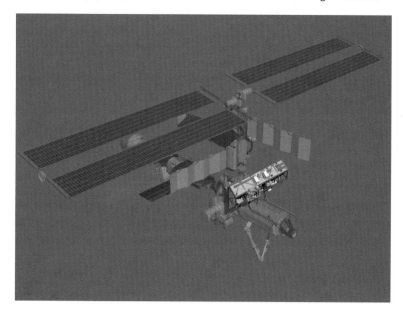

starboard umbilical trays are launched on the forward starboard face.

During the four EVAs to the S0 truss, the astronauts installed first the starboard and then the port struts. The aft strut groups are tripods with three adjustable struts meeting at a common footplate that attaches to Destiny. Each strut has two clamshell fasteners that must be released (for a total of 12 fasteners), and each strut group has five launch restraint bolts that must be released. Six bolts attach each strut group to Destiny's aft end cone and five bolts attach each strut group to the S0 truss. Each clamshell fastener is tightened down to make the struts rigid so that it can accept all loads applied through added truss segments installed on later flights and through movement of the ISS mass in re-boost operations and attitude changes.

Powered subsystems on the S0 truss are cooled through the use of cold plates, which function by running a coolant, ammonia for external hardware, between heat exchanger tubes beneath the Orbital Replacement Unit and a remotely controlled radiator. The Airlock Spur is a structural beam almost 14ft (4.3m) long and fitted with ten EVA handrails, seven long and three short, to allow an astronaut to make his or her way from the Quest airlock across to the forward side of S0 and the Destiny module. It was attached before flight to the starboard aft side of the truss with a hinge attachment and was bolted to the airlock during a spacewalk. In addition, halogen lights were mounted on booms to aid with EVA activities on the night side of the Earth. The first was

installed on Destiny, the second on top of the Unity node near the aft end cone.

Four GPS antennas were preassembled on to the S0 truss to provide the required state vector (position and trajectory) to within 3,000ft (914m) for a single position measurement. Crucial to the enhanced guidance and navigation of the ISS in its growth configuration, two Rate Gyro Assemblies (RGAs) were installed on the aft port side of the S0 truss at the junction between Face 4 and the aft transition face. The RGAs comprise three ring laser gyroscopes, each determining changes of motion in a dedicated co-ordinate axis by measuring changes in the frequency of a reference laser beam using the Doppler effect. The second gyro in the set serves as a redundant back-up. The RGAs allow a second means of measuring the state vector of the ISS in addition to the GPS antennas.

ABOVE LEFT With apparent ease in the weightlessness of space, astronaut Lee Morin moves a keel pin for the S0 truss. *(NASA)*

ABOVE A new GPS antenna for better navigation and position fixing for the station in its continual orbit of the Earth was provided by *Atlantis* on STS-110. *(NASA)*

ANATOMY
Mobile Transporter

Length	9ft (2.74m)
Width	8ft 7in (2.62m)
Height	3ft 2in (0.97m)
Weight	1,950lb (885kg)
Payload capacity	46,100lb (20,910kg)
Speed	0.1–1in/sec

Because remotely controlled operations from inside the ISS call for access to a wide range of locations, the use of Canadarm2 as a mobile, transportable manipulator arm is essential to reducing the workload otherwise conducted

RIGHT As well as
carrying the S0 truss
to the ISS, the April
2002 flight of Atlantis
also saw delivery of
the Mobile Transporter
which was temporarily
parked on the S0
truss. (NASA)

by astronauts on an EVA and for moving large
and bulky packages around the outside of the
station. But Canadarm2 has no fixed base
from which to work and left to its own devices
it moves from place to place hand-over-hand,
which is impractical if it is grasping a package
at one end. For moving between locations and
looking ahead to when the truss segments
would form a continuous path more than 350ft
(107m) long, the Mobile Transporter was lifted
to the ISS by STS-110.

In time it would form an integral part of
the station's robotic systems, providing the
function of a flatbed rail truck to which the
Canadarm2 would be attached by the Mobile
Remote Servicer Base System (MBS), yet to
be installed. In this way Canadarm2, and later
with its special-purpose manipulator called
Dextre attached, the robotic system would do
the work of spacewalking astronauts as well
as performing menial tasks for which an EVA
would be impractical. As such, the MT gave
Canadarm2 the ability to move along the truss
segments on command from an operator
inside the ISS or on the ground. It would also
provide power and data connections between
Canadarm2 and the station elements.

Built of high-strength aluminium, the MT

can relocate Canadarm2 to ten pre-designated
work sites where it can lock itself down with
a 7,000lb (3,175kg) grip for the arm to move
massive payloads. It moves on a three-point
suspension system. The Linear Drive Unit
propels and supports the MT and two Roller
Suspension Units (RSUs) provide additional
stability as it moves down the truss rails. Each
RSU contains two sets of plastic wheels on
spring-loaded mountings. Magnetic sensors on
the MT register metallic strips to tell it when the
MT has reached a traverse stop. Power under
motion is provided by the Trailing Umbilical
System (TUS), a flexible line linking utility
connections to the MT as well as providing data
and communications for commands. Cable
guides are located inside the truss segments
which keep the TUS constrained so as to
avoid snagging protruding structures or fouling
movement of the MT along the rails.

Some 20 different electric motors are needed
to move the MT, latch it down, and plug itself
into a power source, the Umbilical Mechanism
Assembly Ports. Pre-installed on the forward face
(Face 1) of the S0 segment, the MT is the first
moveable robot controlled by software to work
on the outside of a space vehicle, a practical tool
essential to the further build-up of the ISS.

Soyuz TM-34/ISS 4S

25 April 2002, 6.27am GMT
Cdr: Yuri Gidzenko (Russia)
Flight Engineer: Roberto Vittori (ESA)
Tourist: Mark Shuttleworth (US)

The 17th manned mission to the ISS was a Russian 'tourist' flight carrying South African-born Mark Shuttleworth, a highly successful businessman and entrepreneur, accompanied by cosmonaut Yuri Gidzenko and ESA astronaut Roberto Vittori. TM-34 docked to the Zarya nadir port at 7.55am GMT on 27 April. Shuttleworth conducted some biology experiments on board the ISS and carried aboard with him some live rats and sheep stem cells.

This was the last flight of the Soyuz TM variant, with subsequent flights adopting the improved TMA model of the trusty Soyuz. As was standard practice, the tourist mission doubled as a Soyuz replacement flight, the spacecraft being limited to approximately six months in orbit. After switching contour couches to TM-33, and having spent almost eight days aboard the ISS, the three visitors undocked from the Pirs port just after midnight on 5 May and were back on Earth 3hr 21min later after a flight lasting 9 days, 21 hours and 25 minutes.

STS-111/ISS UF-2/Mobile Base System/Expedition 5

Endeavour
5 June 2002, 9.23pm GMT
Cdr: Kenneth D. Cockrell (4)
Pilot: Paul S. Lockhart
MS1: Philippe Perrin (France)
MS2: Franklin Chang-Diaz (6)
MS3 (up): Peggy A. Whitson
MS4 (up): Valery G. Korzun (Russia)
MS5 (up): Sergei Y. Treschev (Russia)
MS3 (down): Carl Walz
MS4 (down): Daniel Bursch
MS5 (down): Yuri I. Onufrienko (Russia)

A logistics resupply, Mobile Base System installation, and Expedition changeover mission, *Endeavour* carried a payload of 36,082lb (16,367kg) including the Leonardo MPLM

LEFT In April 2002, Soyuz TM-34 carried Mark Shuttleworth to the ISS on a 'space tourist' flight arranged by the Russians. *(NASA)*

packed with cargo, the Mobile Remote Servicer Base System, and sundry freight uplifted to the ISS. At two-day intervals, Chang-Diaz and Perrin conducted three spacewalks, the first lasting 7hr 14min for hooking up power and data grapple fixtures to the P6 truss, the second lasting 5hr for attaching the Mobile Base to the Transporter,

BELOW The installation of the Mobile Base to the Mobile Transporter, delivered previously, gave the remote manipulator great flexibility to move around the station on command. *(NASA)*

and the last for 7hr 17min to replace a defunct Canadarm2 wrist joint. This was the last flight for one of France's national astronauts funded by the French space agency CNES, as the remainder of the French astronaut pool was merging with the ESA group. Leonardo carried up eight Resupply Return Stowage Racks and a Resupply Stowage Platform, as well as ExPRESS rack No 3 and a microgravity science glovebox with lithium hydroxide canisters. Various logistics and utility items were also transferred to the ISS. At the end of their flight, which lasted 13days 20hr 35min, they returned to Earth carrying the Expedition 4 crew.

The Expedition 5 crew would support two EVAs, Korzun and Whitson spending 4hr 25min outside on 16 August installing debris impact protection panels on Zvezda, previously parked on PMA-1 before the Russian module docked to Zarya. A second spacewalk was conducted by Korzun and Treschev on 26 August lasting 5hr 21min, during which a space-frame was erected on Zarya ready for further EVAs and new materials supplied by the Japanese Space Agency were installed outside Zvezda. The crew also installed two ham radio antennas on Zvezda as well as a sensor device known as Kromka, designed to measure particulate contamination from that module's attitude control thrusters. The Expedition 5 crew of Korzun, Whitson, and Treschev were retrieved by STS-113 after having been aboard the ISS for 178 days 3hr 33min, and off the Earth for 184 days, 22 hours and 14 minutes.

BELOW A Base System for the Canadarm2 remote manipulator was delivered by *Endeavour* in June 2002, connecting the arm to the Mobile Transporter so that it could run up and down the truss for servicing tasks. *(NASA)*

BELOW RIGHT The Base System was carried to the ISS in the Shuttle cargo bay, installed at the Kennedy Space Center. *(NASA)*

Mobile Remote Servicer Base System (MBS)

Length	18.7ft (5.7m)
Width	14.7ft (4.5m)
Height	9.5ft (2.9m)
Weight	3,307lb (1,500kg)
Handling capacity	46,076lb (20,900kg)
Peak power requirement	825W
Average power requirement	365W

The MBS is essentially a strong and resilient aluminium structure connecting Canadarm2 to the Mobile Transporter, the latter having been delivered to the Z0 truss by the preceding Shuttle flight. The MBS would be the intelligent interface between Canadarm2 and the MT, giving unprecedented mobility to the station's external robotic system. Designed and built for NASA by MD Robotics, the MBS is another ISS element contributed by Canada and expands upon the country's earlier work on the Shuttle arm and on Canadarm2.

Not only does the MBS afford greater mobility to Canadarm2, it also allows the transport of payloads across the station using a fixture called the Payload Orbital Replacement Unit Accommodation (POA). Orbital Replacement Units (ORUs) are packages of subsystems and spares stored on the exterior of the station for retrieval and installation

Canadian Space Agency Agence spatiale canadienne

as replacement parts for worn-out or failed components. The platform's POA would be used primarily for moving large payloads around and for carrying large structures such as truss segments. A secondary device, the MBS Common Attach System, would be employed carrying smaller pallets of works tools and science instruments.

The MBS can also be used as a work platform by astronauts to move along the truss segments or as a place on which to store tools. It doubles as a maintenance platform capable of holding both ends of Canadarm2 should the replacement of any of its joints be necessary. The MBS has four anchor points, Power Grapple Fixtures such as those attached at numerous places around the station by which Canadarm2 can 'walk' from one place to another. These provide electrical power and data pick-off points and connect all the attached robotic systems and their loads to the Robotic Work Station in the ISS.

The MBS also has a single pan and tilt colour camera system located on a mast behind the POA providing a general view of the upper surface and unobstructed views of all four PGFs. Designed for a life of 15 years, the MBS is itself manufactured from a series of separate and interchangeable modules, any one of which can be replaced by an astronaut on EVA or by Dextre once that tool system has been installed and it is equipped with two computers.

STS-112/ISS 9A/S1 Truss/ CETA-1/TRJJ

Atlantis
7 October 2002, 7.46pm GMT
Cdr: Jeffrey S. Ashby (2)
Pilot: Pamela A. Melroy (1)
MS1: David A. Wolff (2)
MS2: Sandra H. Magnus
MS3: Piers J. Sellers
MS4: Fyodor N. Yurchikhin (Russia)

After an absence of four months, during which the Expedition 5 crew were the sole occupants of the ISS, *Atlantis* was again launched to the station. This time it was carrying the S1 truss assembly, a new cooling

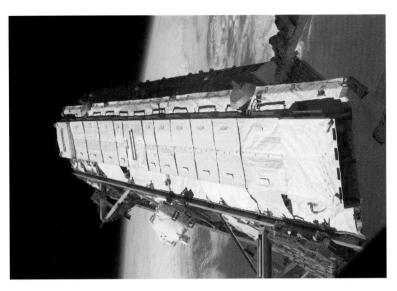

system, an S-band communication system, and the first Thermal Radiator Rotary Joint (TRRJ). Deployment of the S1 truss began when astronauts Magnus and Whitson controlled Canadarm2 from inside the ISS to reach across to the Shuttle payload bay and move it to the starboard side of the S0 truss, attaching it using a claw-like device. After capture, locking bolts structurally mated the two truss segments and only then did Wolff and Sellers exit the Quest airlock to begin their 7hr 1min EVA to lock it firmly in place and begin the lengthy and meticulous task of securing conduits, cables and connectors.

Two days later, on a second EVA lasting 6hr 4min, they conducted installation tasks, setting up a new communication system, radiator modules, fluid lines between the nitrogen tank

ABOVE The S1 truss was lifted across from the Shuttle payload bay to the starboard side of the S0 truss. *(NASA)*

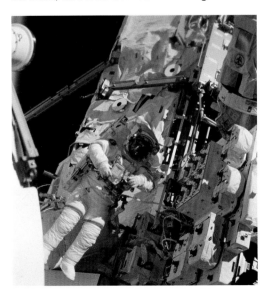

LEFT David Wolff works to fit out the S1 assembly. *(NASA)*

ABOVE Canadarm2 places the Mobile Base on the Mobile Transporter from its launch location in the Shuttle payload. *(NASA)*

BELOW The S1 truss was lifted to the station by *Atlantis* in October 2002. *(NASA)*

ABOVE Canadarm2 attached to the rail-track system along the top of the truss assembly which over time would grow to support eight solar cell array wings. *(NASA)*

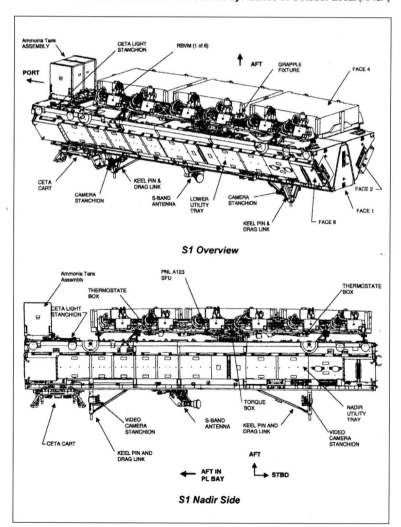

S1 Overview

S1 Nadir Side

assembly and the ammonia tank, and ensuring that seals throughout the truss segments maintained a proper internal pressure. A third and final EVA lasting 6hr 36min allowed completion of jumper connections, removal of port and starboard keel pins, attachment of the last TRRJ connectors, and final securing to the S1.

In addition to the bulk cargo installed on the ISS and 1,604lb (728kg) of water in 16 flexible bags, *Atlantis* returned 1,351lb (613kg) of cargo from the station. Several orbit adjustment and re-boost manoeuvres were completed, leaving the ISS in a 147nm x 219.9nm orbit. Docked for 6 days 21hr 33min, *Atlantis* touched down at the Kennedy Space Center 10 days 19hr 57min after launch.

ANATOMY

S1 and P1 Truss Assemblies

Length	45ft (13.72m)
Diameter	15ft (4.6m)
Height	6ft (1.8m)
Weight	S1 27,676lb (12,554kg); P1 30,871lb (14,003kg)

Construction of the S1 truss began in May 1989 at the Huntington Beach facility, formerly the home of McDonnell Douglas until the company was taken over by Boeing. S1 was delivered to the Kennedy Space Center in October 1999 for flight processing. S1 would be locked into position on the starboard side of Z0, which was bolted down on top of the US Destiny module. Its mirror partner on the ISS would be the P1 truss, launched on 23 November, almost identical but attached to the port side of the Z0 truss. The primary structures comprising bulkheads and longerons are mirrored between the two and Boeing had to design and fabricate different parts. They are all coated with the same optical anodised surface preparations.

One unique difference is that S1 carries an S-band antenna system whereas P1 has a VHF communication system. Both house the Active Thermal Control System which acts like the cooling system in a car radiator except that it uses 99.9 per cent pure ammonia (compared to one per cent in household bleach). The primary structure is made of aluminium and includes seven bulkheads and four longerons per segment, with a heat transport system, trailing umbilical system, on-orbit video camera, electrical equipment, and antenna supports. The upper portion of the radiator assembly rotates to keep the radiators in the shade and there are 18 launch locks that keep this in place, removed by hand during the installation EVA.

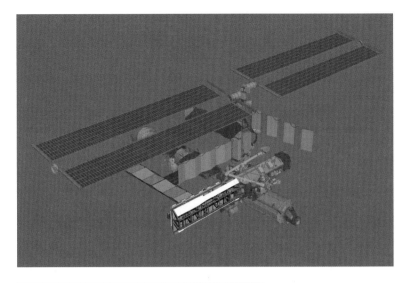

ANATOMY

Crew and Equipment Translation Aid (CETA)

Length	7ft 9in (2.4m)
Width	8ft 3in (2.5m)
Height	2ft 11in (0.9m)
Weight	623lb (283kg)
Volume	468ft³ (13.25m³)

The equivalent of a manually controlled flatbed truck, the CETA consists of a 142lb (64kg) frame incorporating more than 1,100 parts designed to move astronauts and their tools back and forth along the fully assembled truss

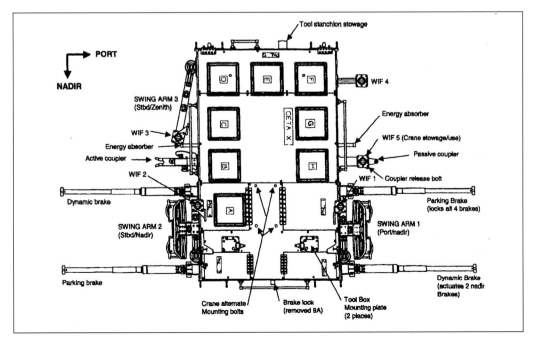

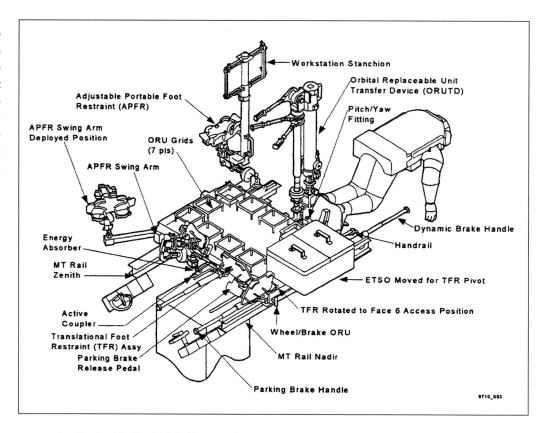

segments attached to the Mobile Transporter
(MT). Two CETAs were flown up to the station,
the first with S1 and the second with P1 truss
assemblies. When installed on the trusses,
the two carts could be located on each side
of the MT – one for each spacewalker –
crewmembers propelling themselves along
on rails using a handbrake system to stop
and start. Almost all the work demanded of
spacewalkers was on and around the truss
assemblies, so the ability to have tool carts and
a powered mobility system significantly eased
the workload during EVA.

If necessary, the two CETA carts could be
placed on the same side of the MT but in all
cases it provided a stable mounting for tools,
equipment, and a variety of attachment points
for transfer devices and packages. When not
used for the astronauts to move themselves
along the truss, the CETA carts would be
carried along with the remotely controlled MT
to which the Canadarm2 would be attached.
In this way, the CETA system expanded the
flexibility of the robotic system, sharing common
equipment for human or robotic activity. Two
CETA toolboxes carried aloft on the S1 were
relocated on to the CETA cart in orbit.

Thermal Radiator Rotary Joint (TRRJ)

Length	3.4ft (1m)
Width	4.5ft (1.4m)
Height	5.2ft (1.6m)
Weight	992lb (450kg)

Also carried aboard STS-112, the TRRJ Orbital
Replacement Unit provides mechanical energy
for rotating the ISS heat rejection radiators
for varying heat rejection rates. Its rotational
capabilities are controlled by the Rotary Joint
Motor Controllers which also power the Drive
Lock Assemblies, which receive signal control
from the rotary joints. One TRRJ is carried
up on each S1 and P1 truss assembly. The
primary structure is aluminium but the drive
and rotating elements are steel, with flex
hoses consisting of corrosion-resistant steel
interfaced with corrosion-resistant steel tubing,
and aluminium quick-disconnects. Several EVA
handholds are provided for spacewalkers to
make electrical and fluid connections during
assembly in orbit.

Soyuz TMA-1/ISS 5S

30 October 2002, 3.11am GMT
Cdr: Sergei Zalyotin (Russia)
Flight Engineer: Yuri Lonchakov (Russia)
Flight Engineer: Frank De Winne (ESA)

Marking the first flight of the improved Soyuz TMA-1 spacecraft, a mixed-nationality crew was launched to the ISS which docked to the Pirs module at 5.01am GMT on 1 November. This short-stay visit was a standard Russian taxi flight to give guest 'astronauts' a ride to the ISS and to exchange a Soyuz spacecraft, and was limited in the time it could remain in orbit. Originally the flight was to have included the musician Lance Bass, who began training in Russia after being financed by sponsors, but they backed out of the deal, leaving Bass without funds. Lonchakov was his replacement. The short-stay crew returned in Soyuz TM-34, the last of the earlier design, leaving TMA-1 at the ISS for the return of the Expedition 6 crew in May 2003. The ISS 5S mission returned to Earth at four minutes past midnight GMT on 10 November after 10 days, 20 hours and 53 minutes.

STS-113/ISS 11A/P1 Truss/CETA-2/TRRJ/ Expedition 6

Endeavour
23 November 2002, 12.50am GMT (November 24)
Cdr: James D. Wetherbee (5)
Pilot: Paul S. Lockhart (1)
MS1: Michael Lopez-Alegria (2)
MS2: John B. Herrington
MS3 (up): Kenneth D. Bowersox (4)
MS4 (up): Nikolai Budarin (Russia)
MS5 (up): Donald R. Pettit
MS3 (down): Sergei Y. Treschev (Russia)
MS4 (down): Valery G. Korzun (Russia)
MS5 (down): Peggy A. Whitson

In addition to exchanging ISS Expedition crewmembers, *Endeavour* delivered the P1 truss segment and a pair of tiny satellites known as MEPSI for the US Department of

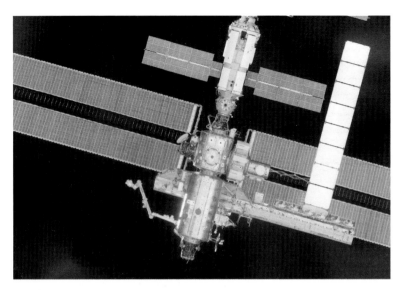

Defense. It also carried the second CETA hand propelled mobility aid and the second Thermal Radiator Rotary Joint. The P1 truss was the mirror image of the S1 truss but carried VHF communications rather than S-band equipment. Riding up for a long stay were Expedition 6 crewmembers Bowersox, Budarin and Pettit. Records were set, some unwelcome. This was the first and only flight for a Native American, John Herrington being an enrolled member of the Chickasaw Nation, and this was the last time a Russian cosmonaut would fly on the Shuttle. It was also the last time a Shuttle would visit the ISS for 32 months and almost four years before the next assembly flight, due to delays made inevitable by a totally unexpected event.

Lopez-Alegria and Herrington conducted three spacewalks at two-day intervals, starting

ABOVE *Endeavour* approaches the ISS where it will install the P1 truss assembly, the S1 truss with its elongated white radiator panel visible from the bottom right. *(NASA)*

BELOW Astronauts Michael Lopez-Alegria and John Herrington install cables and conduits to the P1 truss assembly attached to the port side of the central S0 truss. *(NASA)*

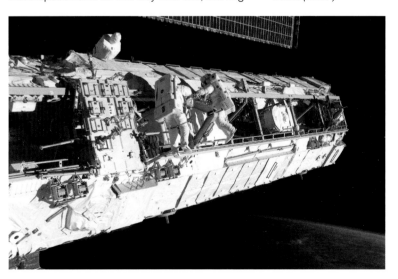

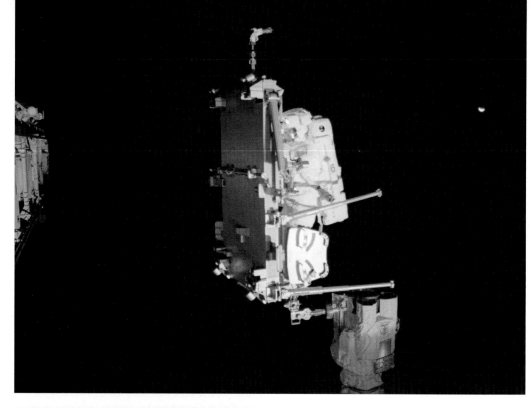

RIGHT On the night side of the Earth, John Herrington moves the second CETA cart to the S1 truss. *(NASA)*

BELOW John Herrington gives scale to the P1 truss and the solar arrays reflecting light from the sun. *(NASA)*

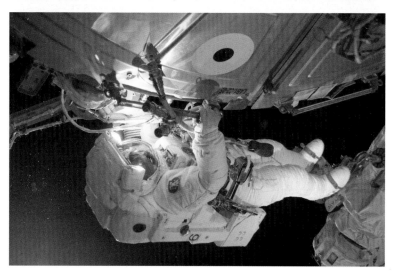

with a 6hr 45min EVA on 26 November to install the P1 truss using techniques similar to those on the deployment of S1. The second EVA lasted 6hr 10min and allowed the spacewalkers to install TV cameras, set up CETA-2 and move it to the S1 truss, and carry out additional installation procedures. The final EVA took exactly seven hours and provided for a check-out of the Mobile Transporter as well as routing electrical harnesses and power bus units. The usual transfers of supplies and equipment took place and after being docked for almost seven days the Expedition 5 crew departed, leaving the three new long-stay tenants aboard the ISS. The crew had expected to land on 4 December but a succession of weather problems held off re-entry for three days and *Endeavour* finally landed at the Kennedy Space Center at an elapsed time of 13 days 18hr 47min.

In space, the Expedition 6 crew set about their long-term stay, which would include two spacewalks by Bowersox and Pettit. The first, on 15 January 2003, lasted 6hr 51min and continued with outfitting the P1 truss assembly with particular emphasis on the radiator assembly, allowing it to be fully

LEFT Piers Sellers works on attaching equipment to the outside of the Destiny module. *(NASA)*

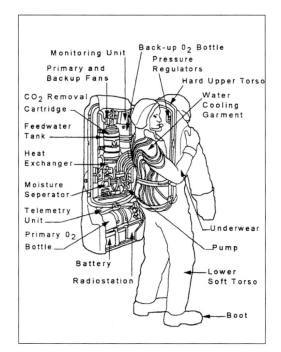

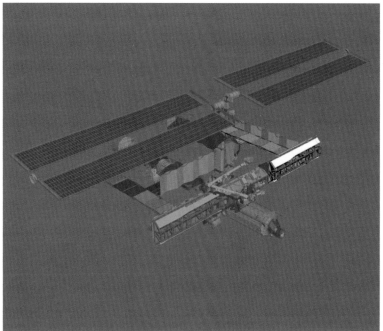

deployed. Other general maintenance duties were conducted which included attending to the ammonia reservoir on the P6 truss and installing a light system on a CETA cart. The second EVA, on 8 April, lasted 6hr 26min and allowed for cables on the S0, S1, and P1 truss assemblies to be reconfigured and for attention to Destiny's heat exchangers.

Expecting to return to Earth aboard the Shuttle, the Expedition 6 crew were in space when *Columbia* was destroyed on 1 February, re-writing the schedules so that Soyuz could be used for Expedition crew changes. Accordingly, having been aboard the ISS for 159 days and 44 minutes, Bowersox, Budarin, and Pettit came home in the Soyuz TMA-1 spacecraft, but a malfunction during descent gave them a rougher and steeper ride than expected. The spacecraft plunged to Earth at Arkalyk, Kazakhstan, at 2.04am GMT on 4 May, 300 miles (483km) off course and with no radio communications after the antenna burned off during re-entry. Their space flight time was 161 days 1hr 14min.

ANATOMY

P1 Truss/CETA-2/TRRJ

Payload details for these elements can be found in the STS-112 payloads description with differences between the two noted in that section.

ANATOMY

MEPSI

The PICOSAT Inspector (MEPSI) programme is funded by the Defense Advanced Research Projects Agency (DARPA) as an experimental research activity into autonomous satellite inspection capabilities. It is under the direction of the US Air Force Research Laboratory in Rome, NY, and it aims to create a technology to image damaged satellites, assess the level of repair work required, and provide feedback to spacecraft operators on the ground. The programme originated with rocket launches of previous satellites in the series during 2000 and 2001. Use of the Shuttle to eject tiny satellites was a great advantage for cost-saving and flexibility.

Built by the Aerospace Corporation and NASA's Jet Propulsion Laboratory, each satellite is 4in x 4in x 5in (10cm x 10cm x 13cm) and when released they operated as a tethered pair ejected from a spring-loaded device on the sidewall of the Shuttle cargo bay at an elapsed time of 8 days 21hr 36min. Joined by a 50ft (15.2m) non-conducting tether, they self-tested a variety of systems and performed measurements on a wide range of fields and electromagnetic interactions. They join a gathering fleet of nanosats incorporating miniaturised components and subsystems packaged into tiny spaces but capable of communicating with the ground.

ABOVELEFT Russian Orlan suits have evolved through experience with Salyut and Mir and are entered through the back, the life support equipment then being closed when the cosmonaut is inside. *(NASA)*

ABOVE The ISS configuration after the installation of the P1 truss assembly on STS-113. *(NASA)*

Chapter Five

ISS Phase Four – final assembly

With anticipation high for an early completion of the giant station, hopes were dashed when the only vehicle capable of transporting big loads into space was grounded. The loss of the Shuttle *Columbia* in 2003 held up assembly for almost four years, while the Russians ferried crews back and forth to the ISS.

OPPOSITE Equipment from the ESP-3 platform is moved across to separate locations on the ISS. *(NASA)*

The next flight after STS-113 was a science mission using a Spacehab double module as a research facility for some highly sensitive scientific experiments. It had nothing to do with the International Space Station and was in an orbit where it could not reach the station under any circumstances. Logically, as it was designated STS-107 it should have flown much earlier, but delays postponed its original flight date and the launch was finally scheduled for January. STS-112 and STS-113 were to have flown after STS-107 but the discovery of major cracks in some Shuttle engines caused a re-scheduling of several flights around the repair work necessary to keep the fleet flying. As a result of this, STS-107 was put back while STS-112 and STS-113 flew their designated missions to the ISS.

During the launch of STS-107 on 16 January 2003, insulation foam from the External Tank to which it was attached broke loose and struck the left wing of the Orbiter *Columbia*, gouging a hole in the leading edge. The flight lasted almost 16 days. When it returned through the atmosphere on the morning of 1 February, hot gases from friction with the air entered the interior of the wing, heating up the structure and causing the complete loss of the Shuttle along with its seven-person crew, which included the first Israeli astronaut. The consequences for the space programme were profound but for the

ISS it emphasised just how much its assembly relied on the Shuttle fleet.

Grounded for 32 months, the station would remain in limbo for almost three years. The exchange of Expedition crews was dependent on Russia's Soyuz spacecraft. Because Soyuz is space-qualified to remain in that hostile environment for no more than six months, this would be the maximum stay time for each crew. Until the disaster, all Expedition recycle flights were to have been conducted by the Shuttle with Soyuz used only for short-stay visits. Now, all that would change, Soyuz recycling the long-stay crews in April and October each year for the two years it took to get the Shuttle fleet flying again. But the consequences for NASA of losing another Shuttle reached beyond the space station.

In the wake of an intensive analysis of the accident, its cause, and the operational viability of the Shuttle programme itself, the review board concluded that the Shuttle should never again fly a mission on which its crew could not reach the ISS, using it as a safe haven until another Shuttle, or a Soyuz, could return the crew. That edict was breached to fly a final repair mission to the Hubble Space Telescope, successfully completed in May 2009. When the Shuttle returned to flight in July 2005 its fate had been sealed, the decision having been made to retire this reusable workhorse after the final assembly of the ISS.

Soyuz TMA-2/ISS 6S/ Expedition 7

26 April 2003, 3.54am GMT
Cdr: Yuri Malenchenko (Russia)
Flight Engineer: Edward Tsang Lu (US)

The Expedition 7 crew were delivered to the ISS by the second of Russia's improved Soyuz spacecraft, arriving at the Zarya nadir port at 5.56am GMT on 28 April. Cosmonaut Yuri Malenchenko and NASA's Edward Tsang Lu joined Expedition 6 crewmembers Bowersox, Budarin, and Pettit, who returned to Earth a few days later on 4 May in the TMA-1 spacecraft. Expedition 7 would remain the resident crew and return in the TMA-3 spacecraft after a mission lasting 184 days 22hr 46min.

BELOW After the loss of Space Shuttle *Columbia* the Russian Soyuz was the only means of reaching the ISS. TMA-2, second in a series of new and improved Soyuz spacecraft, was used to deliver and return to Earth the Expedition 7 crew. *(NASA)*

Soyuz TMA-3/ISS 7S/ Expedition 8

18 October 2003, 5.38am GMT

Cdr: Aleksandr Kaleri (Russia)
Flight Engineer: Michael Foale (US)
Flight Engineer: Pedro Duque (ESA)

The new long-stay crew comprised Aleksandr Kaleri and British-born astronaut Michael Foale, who arrived at the Pirs module at 7.16am GMT on 20 October, but flying with them to the ISS was Pedro Duque of ESA for a brief visit. Duque returned with the Expedition 7 crew in TMA-2 on 27 October. Expedition 8 lasted 192 days 13hr 36min and included one spacewalk, but a significant one nonetheless, in that it was the first time two crewmembers had left the ISS unoccupied for an EVA. One in fact that had to be cut short at 3hr 55min, but not before a variety of experiments and science equipment had been placed around the exterior of the station.

Soyuz TMA-4/ISS 8S/ Expedition 9

19 April 2004, 3.19am GMT

Cdr: Gennady Padalka (Russia)
Flight Engineer: Michael Fincke (US)
Flight Engineer: Andre Kuipers (ESA)

The long-term stay crew of cosmonaut Gennady Padalka and NASA's Michael Fincke were accompanied by ESA astronaut Andre Kuipers making his first flight. He returned with the Expedition 8 crew in TMA-3 on 30 April. Padalka and Foale made four spacewalks during their 185 days, 15 hours and 7 minutes aboard the ISS, totalling 15hr 45min in EVAs conducted on 24 June, 29 June, 3 August, and 3 September. During those spacewalks the crew conducted useful work on only three EVAs, the first being cut at 14 minutes when a pressure problem arose in Fincke's oxygen tank. The other EVAs allowed the crew to carry out

RIGHT An ambassador for European participation, Dutch astronaut Andre Kuipers joined two Expedition 8 crewmembers and returned with Kaleri and Foale in TMA-3. *(NASA)*

LEFT Launched in October 2003, Soyuz TMA-3 delivered the Expedition 8 crew which included British-born astronaut Michael Foale. *(NASA)*

BELOW With dual UK-US citizenship, Michael Foale was the first British astronaut to conduct a space walk. *(NASA)*

essential maintenance work on the station and to prepare hardware for future missions.

Soyuz TMA-5/ISS 9S/ Expedition 10

14 October 2004, 3.06pm GMT
Cdr: Salizhan Sharipov (Russia)
Flight Engineer: Leroy Chiao (US)
Flight Engineer: Yuri Shargin (Russia)

Chiao and Sharipov arrived at 4.16am GMT on 16 October to briefly join Padalka and Fincke, accompanied by cosmonaut Yuri Shargin making his first flight. He would return to Earth with the Expedition 9 crew in TMA-4 on 24 October. Spacewalks on 26 January and 28 March 2005 configured exterior fixtures in anticipation of the arrival of ESA's Automated Transfer Vehicle (ATV) in EVAs totalling 10hr 34min. Expedition 10 had lasted 190 days 19hr 2min when Chiao and Sharipov departed and returned to Earth on 24 April 2005, taking ESA's Roberto Vittori with them.

Soyuz TMA-6/ISS 10S/ Expedition 11

15 April 2005
Cdr: Sergei Krikalev (Russia)
Flight Engineer: John Phillips
Flight Engineer: Roberto Vittori (ESA)

TMA-6 docked to the station on 17 April, returning to Earth in the same spacecraft

on 11 October, completing the bridge between Shuttle flights during the period the fleet was grounded. Vittori returned in TMA-5 but the Expedition 11 crew remained aboard the ISS for 176 days 19hr 30min, and logged space-flight time for this mission of 179 days and 23 minutes. Krikalev was making his sixth space flight and when he landed he had seized the world record for the longest accumulated time in space, a total of 803 days, 9 hours and 39 minutes. The crew conducted a spacewalk on 18 August lasting 4hr 58min during which they removed biological samples from the exterior of the station and replaced materials containers exposed to the vacuum of space.

STS-114/ISS LF-1

Discovery
26 July 2005, 2.39pm GMT
Cdr: Eileen Collins (3)
Pilot: James M. Kelly (1)
MS1: Soichi Noguchi (Japan)
MS2: Stephen K. Robinson (2)
MS3: Andrew S. W. Thomas (3)
MS4: Wendy B. Lawrence (3)
MS5: Charles Camarda

When originally scheduled, *Discovery* was to have flown the ULF-1 mission, a resupply and crew rotation flight, on 1 March 2003, delivering Expedition 7 and returning with the Expedition 6 crew, but that changed after *Columbia* was destroyed on 1 February. Redefined as the LF-1 flight, it would deliver much-needed supplies and logistical stores and bring the Shuttle back into flight operations. Mandated by the review board, the Shuttle was given a second arm. Built by MD Robotics in Canada, this Orbiter Boom Sensor System (OBSS) was stowed down the starboard side of the cargo bay so that the remote manipulator carried as standard on the port side could get an arm extension. After reaching orbit the combined RMS/OBSS would conduct detailed laser scans of areas where the standard arm could not reach, seeking out any blemishes or defects caused by debris during ascent that could imperil the crew.

All three remaining Orbiters were each fitted with 88 impact sensors in the wing leading

BELOW Soyuz TMA-6 displays its rendezvous and docking equipment on the forward face of the Orbital Module. *(NASA)*

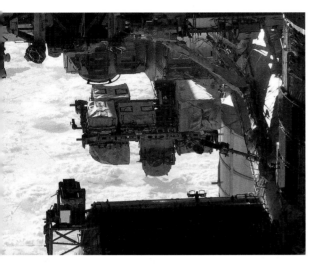

ABOVE With its pallet of replacement units for future use, ESP-2 is attached to the station. *(NASA)*

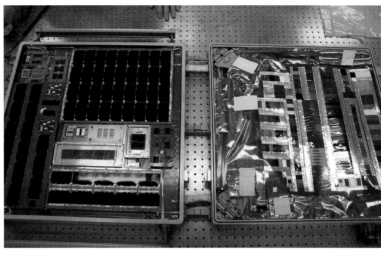

ABOVE Packed in suitcase-size containers, materials exposed to the vacuum of space are transported to the outside of the ISS in MISSE-packets. *(NASA)*

edge connected to laptops on the flight deck registering events that might have damaged the fragile carbon-carbon panels. Moreover, for the first time the ISS would be used as a viewing platform for scanning devices to give the Shuttle a close diagnostic inspection close up and prior to docking. The Orbiter would perform a slow 360-degree back-flip while an astronaut aboard the ISS used high powered telephoto lenses to image the vulnerable thermal protection system. If damage was found there were contingency repair plans for spacewalkers to repair the damage, or the crew could wait in the ISS until another Shuttle was launched to return them.

It had been almost three years since the last Shuttle visited the ISS while four crews, restricted to two people, had come and gone on six-month tenures. When *Discovery* docked with the ISS at 1 day 20hr 53min, it brought much-needed supplies in the Raffaello MPLM and External Stowage Platform 2 (ESP-2) along with a replacement Control Moment Gyro (CMG). More than three months into their stay, the two Expedition 11 crewmembers received their seven Shuttle guests for 8 days 19hr 52min of joint activity, unpacking the MPLM, stowing key items, and supporting three spacewalks by Noguchi and Robinson.

The first spacewalk lasted 6hr 50min during which, on 30 July, Shuttle thermal protection repair procedures were tested, cabling for ESP-2 was installed, and power re-routed to

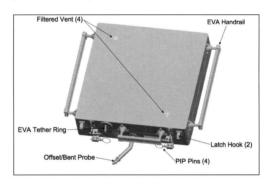

LEFT The MISSE pack has handles for space-walking astronauts to grasp as they emplace and retrieve the experiments it contains. *(NASA)*

CMG-2. The second EVA two days later lasted 7hr 14min. The faulty CMG-1 was removed and installed on a fixture in the Shuttle cargo bay, and the replacement installed in its place. The final EVA took 6hr 1min. It completed installation of ESP-2 on Quest and conducted

BELOW A MISSE experiments package is attached to the top of the P6 solar array, temporarily attached to the Z1 truss. *(NASA)*

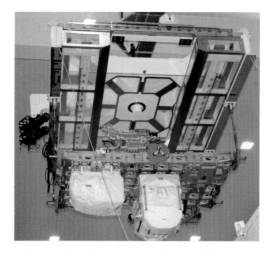

RIGHT Preparing for launch aboard *Discovery* as NASA returns the Shuttle to flight following the loss of *Columbia*, the External Stowage Platform 2 (ESP-2 is installed in a cargo transfer canister. *(NASA)*

an inspection of two suspect thermal tiles on *Discovery* before setting up ham radio equipment. After transferring 3,695lb (1,676kg) of stores to the ISS and accepting 6,600lb (2,994kg) for return to Earth, *Discovery* landed at an elapsed time of 13 days 21hr 32min after some weather delays.

Intensive inspection of the Shuttle's thermal protection system and careful analysis of film and photographic imagery of the launch and ascent phase from cameras on the ground and mounted on the External Tank revealed that foam was still coming away and falling upon the Orbiter during ascent. NASA decided to redesign the external fairings around systems trays running down the side of the ET, thus further delaying ISS assembly while these technical changes were made. Not for another 11 months would Shuttle flights resume, while the Russians hauled replacement crews back and forth to the ISS.

ANATOMY

External Stowage Platform 2 (ESP-2)

Identical in concept to ESP-1 (which was carried to the ISS aboard STS-102), ESP-2 was the second of three carried to space and, like its numerical successor on STS-118,

RIGHT A bridge across the Shuttle cargo bay, the LMC provided a mounting platform to carry equipment that would be transferred to the ISS. *(Boeing)*

FAR RIGHT The mounting base for the External Stowage Platform allowed common attachment points on both the Shuttle and the station. *(Boeing)*

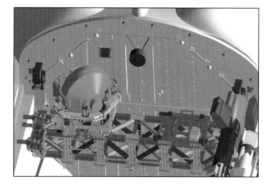

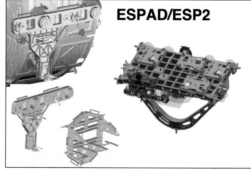

ESPAD/ESP2

RIGHT Upper and lower views of the External Stowage Platform carried to the ISS by *Endeavour*. *(Boeing)*

Empty FRAM   VSSA

Empty FRAM

UTA   FHRC

Empty FRAM

ESPAD   MBSU

Empty FRAM

ESP2 LAUNCH CONFIGURATION (TOP VIEW)

ESPAD

ESP-2

ESP2 LAUNCH CONFIGURATION (BOTTOM VIEW)

very much bigger. With a length of 8ft (2.4m) and 13ft (4m) wide, it carried seven ORUs (Orbital Replacement Units) out of a possible eight locations and included two direct current switching units, two main bus switch units, a pitch/roll joint, a flexible hose rotary coupler, and a utility transfer assembly. Cantilevered off the side of the Quest airlock, ESP-2 was, like ESP-1, powered by Unity. Various ORUs were moved on and off the platform as assembly and operations with the ISS evolved.

Soyuz TMA-7/ISS 11S/ Expedition 12

1 October 2005, 3.55am GMT
Cdr: Valery Tokarev (Russia)
Flight Engineer: William S. McArthur (US)
Tourist: Gregory Olsen (US)

TMA-7 was an ISS Soyuz exchange mission delivering the next long-duration crew, docking with the Pirs module two days later. Their spacecraft would remain for six months and in that time be relocated first to the nadir port on Zarya on 18 November and then to the aft port of Zvezda on 20 March before returning back to Earth on 8 April. Space tourist Gregory Olsen spent eight days participating in experiments before returning to Earth with the Expedition 11 crew on 10 October. There was to have been a third crewmember to Expedition 12 but ESA astronaut Thomas Reiter would have to wait while further delays to the Shuttle schedule deferred his ride to Expedition 13, opening the seat slot for Olsen.

Tokarev and McArthur conducted two spacewalks, the first lasting 5hr 22min on 7 November, to install a new camera on P1 and to jettison a failed instrument. The second was on 3 February 2006 and lasted 5hr 43min, during which they jettisoned an old Orlan suit equipped for broadcasting a radio signal to amateur hams on the ground, and retrieved some experiments. While unaccustomed to celebrity wake-up calls, the crew received a special start to their day on 3 November when Sir Paul McCartney broadcast a live performance of 'Good Day Sunshine' and 'English Tea' during a live concert in Anaheim, California. The crew departed on 8 April 2006 in TMA-7 after a stay of 187 days 14hr 1min with

a total space-flight time of 189days 19hr 53min. They were relieved by the Expedition 13 team that had arrived a week earlier.

This was the last of the 11 Soyuz missions for conveying US and Russian crews to the ISS, signed as part of the 1996 agreement for space station participation. Additional flights were necessary and further negotiations took place under a political cloud that required special dispensation from the Iran Non Proliferation Act signed in 2000.

Soyuz TMA-8/ISS 12S/ Expedition 13

30 March 2006, 2.30am GMT
Cdr: Pavel Vinogradov (Russia)
Flight Engineer: Jeffrey Williams (US)
Science Officer: Marcos Pontes (Brazil)

TMA-8 docked to the Zarya nadir port at 4.19am GMT on 1 April. They were accompanied on the flight up by Brazilian test pilot Marcos Pontes, who had been working at the NASA Johnson Space Center on his country's contribution to the ISS and was making his first space flight. He had been training in Houston for a NASA Shuttle flight but that was transferred to Russia when the Shuttle encountered scheduling problems. After more than seven days aboard the ISS, Pontes returned to Earth with the Expedition 12 crew

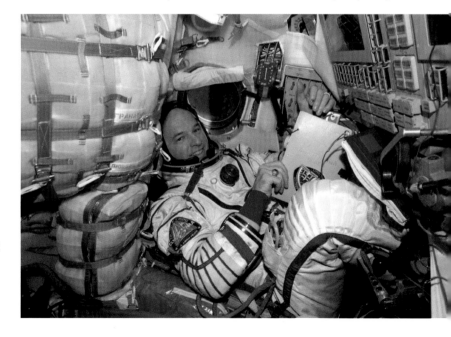

BELOW With little room for movement, NASA's Jeffrey Williams finds life aboard Soyuz TMA-8 a little cramped! *(NASA)*

aboard TMA-7, departing the aft port on Zvezda at 8.28pm GMT on 8 April.

Thomas Reiter became part of the Expedition 13 crew when he arrived aboard *Discovery* on the STS-121 mission on 6 July, restoring ISS crew size to three for the first time since May 2003. He became part of the Expedition 14 crew when Lopez-Alegria and Tyurin were launched aboard TMA-9 in September 2006 and returned to Earth aboard STS-116 on 22 December. Vinogradov and Williams had been aboard the ISS for 180 days 17hr 34min.

During the mission three crewmembers performed two spacewalks. The first, on 1 June, lasted 6hr 31min, during which Vinogradov and Williams carried out some systems repairs to the Elektron oxygen production unit and retrieved a selection of experiment packages. The second EVA was conducted on 3 August and in 5hr 54min Williams and Reiter installed materials processing experiments outside the station, installed a Thermal Radiator Rotary Joint (TRRJ) control unit, and performed sundry maintenance and servicing tasks.

Vinogradov and Williams returned to Earth in TMA-8 with short-stay visitor Anousheh Ansari, separating from the nadir port on Zarya at 9.53pm GMT on 28 September and landing at 1.13am GMT on 29 September. Expedition 13 had been at the ISS for 180 days 17hr 34min, and the two long-stay crewmembers had logged just 16 minutes short of 183 days from launch to landing.

STS-121/ISS ULF1.1

Discovery
4 July 2006, 6.38pm GMT

Cdr: Steven W. Lindsey (3)
Pilot: Mark E. Kelly (1)
MS1: Michael E. Fossum
MS2: Lisa M. Nowak
MS3: Stephanie D. Wilson
MS4: Piers J. Sellers (1)
MS5 (up): Thomas Reiter (ESA)

In a flight as much about qualifying the redesigned Shuttle External Tank as about resuming operations to the ISS, *Discovery* lifted the Leonardo MPLM to the station with further quantities of supplies and stores still very much in demand due to the Shuttle delays. *Discovery*

docked with PMA-2 less than 45 hours after launch, and the seven astronauts were welcomed by the two Expedition 13 crewmembers now in their fourth month aboard the ISS.

Leonardo was lifted across to the Unity module before the first EVA began from the Quest airlock on the fourth day, a spacewalk which lasted 7hr 31min and tested the OBSS arm as a potential work platform for inaccessible and remote places on the ISS. In tests using this attached to the Shuttle arm, both astronauts evaluated the effectiveness of being moved around a variety of difficult-to-reach exterior locations. Fossum and Sellers also worked on the Mobile Transporters and routed a special trailing cable which had shown signs of failure.

Conducted two days later, the second EVA consisted of installing a spare pump module, conducting further maintenance tasks, and transferring equipment to the ESP-2 in activity that lasted 6hr 47min. The third EVA two days later consisted of a busy set of detailed objectives involving the thermal control repair kits and other sundry items. In all, 7,423lb (3,367kg) of supplies and stores was lifted from Leonardo and placed aboard the ISS with a further 1,863lb (845kg) from the middeck and 1,546lb (700kg) of water. *Discovery* received 6,451lb (2,926kg) from the ISS, mostly stored in the MPLM.

With two resupply flights completed and qualification of the design changes to the External Tank to minimise foam hazards, the Shuttle could resume ISS assembly missions after a period of almost four years during which the physical size of the station had not grown.

Discovery landed back at the Kennedy Space Center at an elapsed time of 12 days 18hr 37min with Thomas Reiter left behind on the ISS to join the long-duration crew of Expedition 13 to return aboard STS-116 in December.

ANATOMY

Minus Eighty Degree Laboratory Freezer (MELFI)

Developed by the European Space Agency and built by a consortium led by EADS-Astrium, MELFI weighs 1,609lb (730kg) and is a rack-sized facility which will provide the station with refrigeration for life science and biological samples. MELFI comprises four separate storage vessels adjustable to different

temperatures, and while it has an operating range of between +10°C and –99°C, the most commonly used temperature settings aboard the ISS are –80°C, –26°C, and –4°C. The cooling system is a high-tech solution incorporating a sophisticated technology that uses less than 1kW to operate.

Mounted within a cold box, the two heat exchangers consist of a total of 7 miles (11.25km) of piping. The cooling fluid is nitrogen and the entire cooling system is an Orbital Replacement Unit that can be changed within

eight hours, during which the insulation will preserve the pre-set temperature. With a total capacity of 300 litres, the first MELFI was carried to the ISS by the STS-121 mission in 2006 and installed in the US Destiny module under the guiding hand of European astronaut Thomas Reiter. Two MELFI units were supplied to the Americans and a third to the Japanese. The last MELFI was carried up to the ISS aboard *Discovery* on the STS-131 mission in 2010. Each MELFI is housed in a MPLM, also provided by the Europeans.

ABOVE LEFT Gregory Chamitoff works the European MELFI refrigeration facility for biology experiments aboard the ISS. *(NASA)*

ABOVE Sunita Williams carefully inserts blood samples into the MELFI freezer for analysis on board the station. *(NASA)*

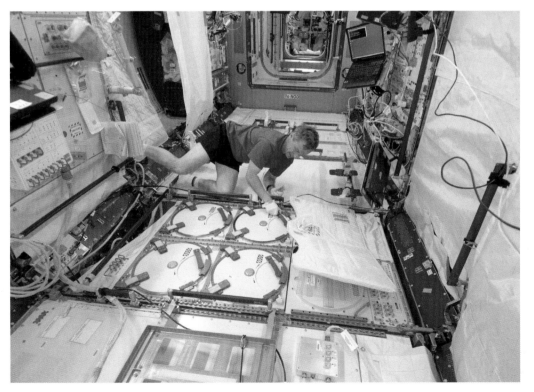

LEFT Robert Thirsk works the MELFI equipment. Three such facilities were provided for US, European and Japanese segments of the ISS. *(NASA)*

Oxygen Generation System (OGS)

Transported to the ISS in the Leonardo MPLM, the 1,465lb (665kg) OGS is the first of two core components of the oxygen generation system for the regenerative ECLSS. Using water to generate breathable oxygen, it was the first step towards providing an environmental support system for six crewmembers, the optimum size for station crews. Designed and built by NASA Marshall Space Flight Center and Sundstrand Space Systems International, it replaces oxygen lost during repressurisation. In a related and highly valuable way, the OGS is a transition to the form of oxygen generation system that would be needed for very long flights, such as those from Earth to Mars and the only way future long-duration missions could be supported.

In operation, the OGS can produce up to 20lb (9kg) of oxygen per day, more than the 12lb (5.4kg) required for a six-man crew, and uses the station water supply to split the liquid into hydrogen and oxygen molecules. The hydrogen is vented to space and the oxygen is directed into the atmosphere. Integral to the OGS, the water recovery system launched on STS-126 in November 2008 produces clean and usable liquid from recycled waste and urine, with a far greater purity than average tap water on Earth. The two sides of the system – OGS and water recovery – are packaged into refrigerator-sized racks inside Destiny.

BELOW *Atlantis* **docks to PMA-2 at the forward end of the US Destiny module, the P1 truss visible in the foreground.** *(NASA)*

STS-115/ISS 12A/P3-P4 Integrated Truss Segment

Atlantis
9 September 2006, 3.15pm GMT

Cdr: Brent W. Jett (3)
Pilot: Christopher J. Ferguson
MS1: Joseph R. Tanner (3)
MS2: Daniel C. Burbank (1)
MS3: Heidemarie M. Stefanyshyn-Piper
MS4: Steven G. MacLean (1) (Canada)

It had been nearly four years since the last assembly flight and when *Atlantis* lifted off at 3.15pm GMT on 21 September, it was with a great sense of relief that both Shuttle and ISS build-up was on track, a feeling that the station was now on the home run to completion. But there was still a long way to go, with several truss assemblies and pressurised modules from Europe and Japan yet to be installed. After docking with PMA-2 at 11.01am on 11 September, the six-person crew joined the three Expedition 13 crewmembers to begin a busy six days involving installation of the P3/P4 truss segment and three spacewalks. Ferguson and Burbank used the Shuttle's remote manipulator arm to grasp the truss assembly and lift it across to Canadarm2 on the ISS operated by MacLean and Williams.

With the P3/P4 truss fixed to P1, on 12 September Tanner and Stefanyshyn-Piper began the first spacewalk during which they secured the truss and got so far ahead of the schedule that they began activities planned for the second EVA. After the first 6hr 26min EVA, Burbank and MacLean prepared to go outside the following day and in a 7hr 11min spacewalk they prepared the Solar Array Alpha Rotary Joint (SARJ) rotation assembly for activation, the first of two that would be installed on the ISS, each at opposing ends of the central truss structure.

On 15 September the final EVA lasted 6hr 42min, during which Tanner and Stefanyshyn-Piper tidied up the assembly site, unfurled the PVR radiator, replaced an S-band antenna, and prepared the area for the next work crew coming up in three months. Activation of the solar arrays would await the next Shuttle mission for a major wiring job to connect the

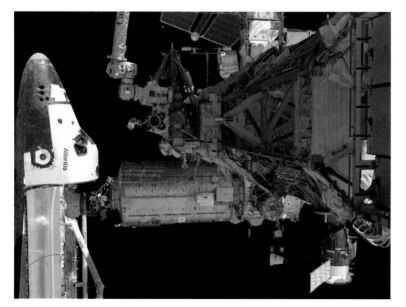

panels to the ISS electrical system. After the EVA the Mobile Transporter was moved to the P3 truss ready for the next tasks. After 6 days 2hr 4min at the station, *Atlantis* undocked and returned to the Kennedy Space Center on 21 September at 10.21am GMT, 48 minutes before sunrise local time, a mission time of 11 days 19hr 7min.

ANATOMY

P3/P4 Integrated Truss Segment

Length	45.3ft (13.8m)
Width	16ft (4.9m)
Height	15.6ft (4.8m)
Weight	34,885lb (15,824kg)

With the S0 segment as the central attachment point to Destiny, the elongated truss structure designed to support eight sets of solar arrays was built outwards in segments, numbered according to spacing with 'P' and 'S' denoting port or starboard locations. Originally there were to have been sections designated P1 to P6 and S1 to S6 but under the original design P2 and S2 were to have incorporated thrusters for attitude control. These were deleted when the station was redesigned, leaving only five each side of the Z0 truss segment. The P6 truss and solar array assembly was temporarily attached to the Z1 section so as to supplement electrical power production from the Russian segment of the ISS, but this would eventually be relocated to its final position as the outermost segment of the port side of the truss. The P3/P4 assemblies were mated to the P1 segment carried to the ISS by *Endeavour* on STS-113.

The P3/P4 segment is a hexagonal aluminium structure with four bulkheads and six longerons incorporating brackets, fittings, platforms, EVA equipment, and miscellaneous mechanisms in a secondary structure. Major elements on the truss are the attachment system for mating it to the P1 segment, a Solar Array Alpha Rotary Joint (SARJ) and an Unpressurised Cargo Carrier Attach System (UCCAS). Prime functions for the truss are mechanical, power, and data interfaces to payloads attached to the two UCCAS platforms, axial tracking of the sun for the solar arrays, and work site accommodation for the

Mobile Transporter. There is also a conduit for transferring ammonia coolant to the outboard photovoltaic modules plus two multiplexer/demultiplexer units. The UCCAS doubles as a stable platform for payloads and ORUs and provides power and data to these units.

Pre-mated during manufacture, the P3 and P4 truss assemblies are quite different and carry separate items of equipment. P3 consists of the SARJ which keeps the Solar Array Wings (SAWs) on P4 and P6 orientated towards the sun as the station orbits the Earth. This combination works like a throttle control on a motorcycle, where P3 is the fixed handlebar and P4 is the twist-grip, with the pivot for the grip located inside the handlebar. P4 carried the two folded Solar Array Wings deployed only after

ABOVE Attached to the P1 truss, the dual P3/P4 unit displays the two opposing solar array wings yet to be deployed from their T-bar containers (right). *(NASA)*

BELOW Up and running, the P4 array unfurled from its concertina-packaged deployment box. *(NASA)*

the P3/P4 assembly had been attached to the P1 truss. P4 is the second of four Photovoltaic Modules (PVMs) assembled at the station, the first being P6 launched by STS-97 in November 2000 and is similar to that. Details of the Solar Array Wings can be found under that mission's payload description.

Built by Lockheed Martin, the SARJ unit attached to P3 was the first of two that would be carried to the ISS. It would continuously rotate to keep the SAWs on P4 and (eventually) P6 aligned with the sun through a short intermediate segment, P5. The SARJ is a 10ft (3m) diameter rotary joint, or wheel, that weighs 2,500lb (1,134kg) and rotates on bearing assemblies powered by a servo control system. It is the linking interface between P3 and P4. Power flows through a utility assembly with roll-ring power transfer. The Beta Gimbal Assembly

(BGA), measuring 3ft x 3ft x 3ft (91cm x 91cm x 91cm) is the same as that carried on P6 and changes the pitch of the wings by slowly spinning the solar array. Both the SARJ and the BGA are pointing mechanisms, following an angled target – the sun – rotating on two axes to always point at the source of solar energy. The SARJ mechanism rotates 360 degrees every orbit as the station circles the Earth at a fixed attitude in relation to the surface, equivalent to about 4 degrees per minute, and to do that it rotates the entire P4 truss element.

P3 was designed by the Boeing facility at Huntington Beach and P4 was developed at the former Rocketdyne Power and Propulsion facility at Canoga Park, California, now a division of Pratt & Whitney. Assembly of P3/P4 took place at Tulsa, Oklahoma, in 1997. Like the P6 truss, the P3/P4 array has a Battery Charge/Discharge Unit (BCDU) and a Photovoltaic Radiator (PVR) for which details can be found at the P6 payload description.

Soyuz TMA-9/ISS 13S/ Expedition 14

18 September 2006, 4.08am GMT
Cdr: Mikhail Tyurin (Russia)
Flight Engineer: Michael Lopez-Alegria (US)
Tourist: Anousheh Ansari (US)

Launched just hours before *Atlantis* departed the ISS, the two-man Expedition 14 crew were launched from Baikonur accompanied by

Iranian-born Anousheh Ansari. Co-founder of Telecom Technologies in the US, she was the fourth space tourist and the first self-funded woman astronaut. Her family co-sponsored the Ansari-X prize, a competition to fly the first commercial human space flight to the edge of space twice within two weeks. That prize was won on 4 October 2004 when the Scaled Composites SpaceShip One rocketed to an altitude of just over 100km, for record breaking purposes the arbitrary boundary between Earth's atmosphere and space. TMA-9 docked to the aft port on Zvezda at 5.21am GMT on 20 September and the visitors joined the three crewmembers of Expedition 13, including Thomas Reiter, who would remain and become the third crewmember of Expedition 14. During their tenure Lopez-Alegria and Tyurin were joined by Sunita Williams, flown up aboard *Discovery* in December 2006 to replace Reiter.

Expedition 14 would conduct five spacewalks totalling 33hr 42min. During the first spacewalk on 22 November Tyurin and Lopez-Alegria prepared equipment for Europe's Automated Transfer Vehicle. On the second EVA, conducted on 31 January 2007, Lopez-Alegria and Sunita Williams (who had by now replaced Reiter) reconfigured two coolant pumps on Destiny and carried out general maintenance work. During the third EVA on 4 February Lopez-Alegria and Williams completed work on an ammonia servicing panel and took images of the P6 starboard solar wing before its retraction prior to the arrival of STS-117. On 8 February, during the fourth EVA involving Lopez-Alegria and Williams, shrouds were removed from the rotary control motors on P3 and various cables were attached to PMA-2. The final EVA for this crew was performed by Lopez-Alegria and Tyurin on 22 February when further servicing work was performed, this time to the Russian sector.

Anousheh Ansari returned to Earth with Vinogradov and Williams in TMA-8, landing at 1.13am GMT on 29 September, her flight having lasted 10 days 21hr 5min. Expedition 14 crewmembers Tyurin and Lopez-Alegria would return to Earth aboard TMA-9 on 21 April 2009, having been in space for 215 days, 8 hours and 22 minutes, an unusually long period for a Soyuz to remain in orbit.

STS-116/ISS 12A.1/P5 Truss Segment

Discovery
9 December 2006, 1.48am GMT (10 December)
Cdr: Mark L. Polansky (1)
Pilot: William A. Oefelein
MS1: Nicholas J. M. Patrick (1)
MS2: Robert L. Curbeam Jr (2)
MS3: Christer Fuglesang (ESA)
MS4: Joan E. Higginbotham
MS5 (up): Sunita L. Williams
MS5 (down): Thomas Reiter

When *Discovery* launched from the Kennedy Space Center at 8.48pm local time (1.48am GMT on 10 December) it marked a resumption of Shuttle-based expeditionary crew deliveries, suspended since the loss of *Columbia* almost four years before. But it was only to drop off Sunita Williams, joining Expedition 14 now three months in to their stay, and to return with Thomas Reiter, who had bridged both this and the preceding long-duration crews. Going up was a cargo load of 35,690lb (16,189kg) comprising the P5 Integrated Truss Segment, an 11,900lb (5,398kg) pressurised Logistics Single Module

LEFT The payload bay doors are closed over the Spacehab logistics module (top) and the P5 truss segment. *(NASA)*

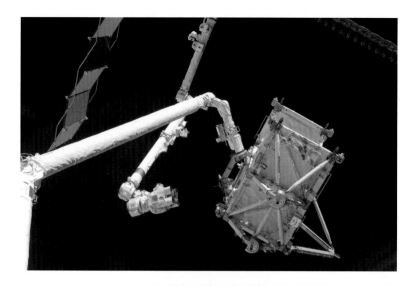

ABOVE P5 gets a handoff from the Shuttle manipulator arm to Canadarm2 on the ISS. (NASA)

RIGHT Resistant to efforts at retracting the P6 solar array the thin solar cell arrays prove difficult to fold back up. (NASA)

(LSM) connected by a 250lb (113kg) tunnel to the Shuttle middeck, and a 6,490lb (2,944kg) Integrated Cargo Carrier (ICC) with multiple packages of equipment for transfer to the ISS.

Discovery docked to the ISS at 10.26pm on 11 December and in a 6hr 36min spacewalk the following day, Curbeam and Fuglesang directed the final mating of the P5 to the P4 truss segment, installing power and heater cables as well as replacing a faulty camera on the S1 truss. Before the SARJ unit on the P3/P4 truss could be activated the port side Solar Array Wings on P6 (4B) had to be retracted, and on 13 December that was attempted, but with only moderate success. An apparent loss of tension in the guide wires produced kinks in the arrays and through a series of staggered attempts only 14 of the 31 bays could be drawn down, leaving 17 still out. Nevertheless, that was sufficient clearance to start up the P3/P4 arrays tracking the sun for the first time through the SARJ wheel. Next day, 14 December, the second EVA lasted an even five hours when Curbeam and Fuglesang began to wire up the station's permanent electrical systems, and electrical power was switched on through the P4 arrays for the first time.

The third EVA took place on 16 December and lasted 7hr 31min during which great efforts were placed on trying to get more of the P6 4B arrays retracted, but still 11 remained out.

RIGHT The Integrated Cargo Carrier installed in Discovery as a mount for small items transferred to the ISS. (NASA)

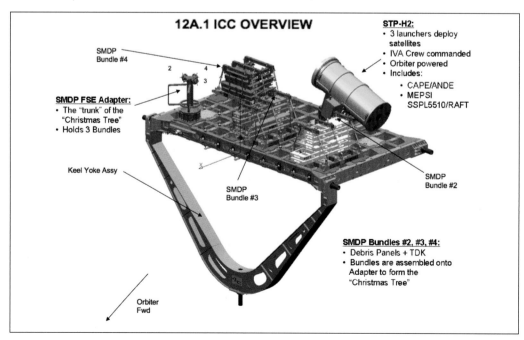

12A.1 ICC OVERVIEW

STP-H2:
- 3 launchers deploy satellites
- IVA Crew commanded
- Orbiter powered
- Includes:
  - CAPE/ANDE
  - MEPSI
    SSPL5510/RAFT

SMDP Bundle #4

SMDP FSE Adapter:
- The "trunk" of the "Christmas Tree"
- Holds 3 Bundles

Keel Yoke Assy

SMDP Bundle #3

SMDP Bundle #2

SMDP Bundles #2, #3, #4:
- Debris Panels + TDK
- Bundles are assembled onto Adapter to form the "Christmas Tree"

Orbiter Fwd

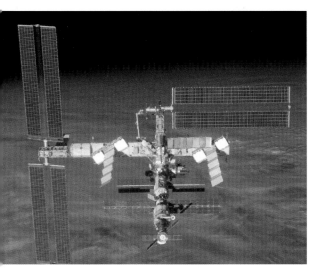

With work still to do, on 17 December flight controllers extended the Shuttle mission by one day for a fourth EVA. Conducted on 18 December and lasting 6hr 38min, the remaining SAW wing was retracted the last 36ft through a series of shaking moves using the guide wires and staggered retract commands. On 20 December, the crew aboard *Discovery* released two MEPSI nanosats and two 5in² Radar Fence Transponder satellites for the Navy and the following day they released two ANDE microsatellites for the Naval Research Laboratory. Leaving Sunita Williams as a late third crewmember for Expedition 14 and returning on 22 December with Thomas Reiter, *Atlantis* landed at an elapsed mission time of 12 days 20hr 44min.

ANATOMY

Spacehab Logistics Single Module

Length	10ft (3m)
Width	14ft (4.3m)
Height	11ft (3.4m)
Payload capacity	6,000lb (2,720kg)
Pressurised volume	1,000ft³ (28.32m³)

The SLSM is a pressurised aluminium cylinder designed to provide storage and work space for crewmembers, and is connected to the Shuttle middeck by a tunnel. It can accommodate a maximum of 118 Shuttle middeck locker equivalent volumes on the bulkheads, the double rack, and the special stowage system.

ANATOMY

P5 Integrated Truss Segment

Length	11ft 1in (3.4m)
Width	14ft 11in (4.5m)
Height	13ft 11in (4.2m)
Weight	4,110lb (1,864kg)

The P5 segment serves as a spacer between the P4 and P6 truss and solar array assemblies on the port end of the 11-part integrated truss structure. When delivered to the ISS the P6 array had been parked on the Z1 truss, but with the spacer delivered by STS-116 there was now a location to which it could be transferred to support the P4 array on that end of the ISS backbone. The former Rocketdyne (now Pratt & Whitney) company had designed and built P5, which had been delivered to the Kennedy Space Center in July 2001 – its launch, however, was delayed by the loss of *Columbia* and the grounding of Shuttle vehicles. Fabricated primarily from aluminium, in addition to serving as a locating segment for the P6 array, P5 provides numerous locations for external stowage and allows platforms to be attached for ORUs, science packages, or thermal blankets for shading the Solar Array Assemblies (SAAs).

P5 was installed on the P4 segment by Canadarm2, which had been removed from the Shuttle's cargo bay by the Orbiter RMS and handed over to the station's manipulator arm. A special grapple device on P5 was used by each arm to handle the segment as it moved carefully

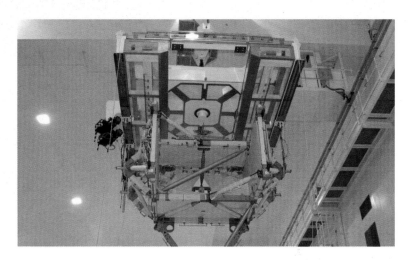

across to the ISS. This fixture was later removed and stowed on the truss keel. The P5 was installed robotically but with the crew assisting and monitoring, a delicate task given that there was a mere 3in (7.6cm) clearance on the P4 Sequential Shunt Unit. Four 0.75in (1.9cm) diameter primary bolts in each corner of the segment were driven home during the first spacewalk with contingency bolts also provided. In addition to supporting external packages, the truss also had an External Wireless Instrumentation System (EWIS), a package of sensors designed to monitor the modal response of the truss assembly to vibration and forces induced by attitude changes and re-boost activity.

Soyuz TMA-10/ISS 14S/ Expedition 15

7 April 2007, 5.31pm GMT
Cdr: Oleg Kotov (Russia)
Flight Engineer: Fyodor Yurchikhin (Russia)
Tourist: Charles Simonyi

An expeditionary crew changeover taxi flight, TMA-10 docked to the ISS at 10.10pm GMT. The crewmembers joined Tyurin, Lopez-Alegria, and Williams, the latter having arrived at the station on 11 December, where she would remain until returning aboard STS-117 in June. Originally expecting to return with STS-118, on 26 April NASA decided to bring her back on STS-117 when hail damage to that mission's External Tank delayed the flight with the consequent knock-on effect for its successor.

Expedition 15 was left alone when TMA-9 undocked from Zarya's nadir port at 9.11am

GMT on 21 April and returned to Earth with Expedition 14 crewmembers Tyurin and Lopez-Alegria at 12.31pm GMT that day. They returned with Simonyi after his 13 day 19hr mission, one day longer than planned due to unsuitable conditions on the ground. TMA-9 had been in space for 215 days, five days longer than the certificated on-orbit duration for the Soyuz spacecraft.

During their mission, Yurchikhin and Kotov conducted three spacewalks. The first, on 30 May, lasted 5hr 25min, to install debris protection panels around the Zvezda module and rewire a GPS antenna. The second, on 6 June, lasted 5hr 37min and saw the spacewalkers lay out an Ethernet on Zarya and install more impact protection panels in addition to deploying some experiments. A third EVA took place on 23 July, by which time Clayton Anderson had replaced Williams aboard STS-117, and he and Yurchikhin spent 7hr 41min replacing some equipment on the Mobile Transporter and doing general maintenance duties.

Kotov and Yurchikhin undocked from the ISS at 7.14am GMT on 21 October, returning with space tourist Sheikh Muszaphar Shukor. The de-orbit burn fired at 9.47am but a fault in the Soyuz guidance system similar to that which affected the re-entry of TMA-1 tipped the spacecraft into a high-g ballistic trajectory and TMA-10 landed at 10.46pm GMT, 210 miles (338km) off target, west of Arkalyk. The same problem would occur again with TMA-11. Expedition 15 had logged 196 days, 17 hours and 17 minutes in space.

STS-117/ISS 13A/S3-S4 Integrated Truss Segment

Atlantis
8 June 2007, 11.38pm GMT
Cdr: Frederick W. Sturckow (2)
Pilot: Lee J. Archambault
MS1: Patrick G. Forrester (1)
MS2: Steven R. Swanson
MS3: John D. Olivas
MS4: James F. Reilly II (2)
MS5 (up): Clayton C. Anderson
MS5 (down): Sunita L. Williams

In several respects, the schedule for *Atlantis* was a repeat of the flight that brought up the P3/P4 arrays nine months earlier.

Forrester, Swanson, Olivas, and Reilly would perform spacewalks while Anderson would join Expedition 15 aboard the ISS, relieving Williams. Originally planned to fly on 15 March, *Atlantis* was late getting off the pad due to hailstorm damage to its External Tank. Docking to the PMA-2 port occurred at 7.36pm on 10 June, and the hatches to the ISS were opened 90 minutes later.

While the S3/S4 truss was lifted across from the Shuttle cargo bay as planned, activation of the SARJ unit brought problems, and retraction of the P6 2B solar array to stow it out of the way had difficulties similar to those with array 4B on the preceding flight. Then the central and terminal computers in the Russian segment failed, and were restored with jumper cables and bypass devices. Four spacewalks were necessary totalling 27hr 59min, during which the P6 arrays were totally retracted – awaiting the time they would be moved to the port end of the truss assembly – and the SARJ mechanism was extensively documented for analysis.

A total payload of 42,641lb (19,342kg) at

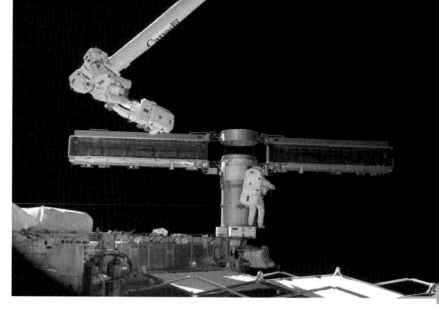
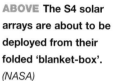

launch made this one of the heaviest Shuttle loads but most of that was the P3/P4 truss assembly. For the first time, the asymmetric configuration of the ISS due to the temporary location of the P6 solar arrays had been changed into a layout more recognisable as the final plan. After 8 days 19hr 6min attached to the ISS, *Atlantis* undocked and ended at a mission time of 13 days 20hr 12min, a landing delayed one day by bad weather, returning

ABOVE The S4 solar arrays are about to be deployed from their folded 'blanket-box'. *(NASA)*

FAR LEFT The fixed (upper) S3 truss attached to the S4 section carrying the solar array wings folded for launch. Note the folded radiator panels to the left of the S4 section. *(NASA)*

LEFT The open framework of the hexagonal S3 truss is clearly visible and displays routes by which a space walking astronaut can get inside for installation. *(NASA)*

RIGHT With the P6 arrays retracted in their temporary location above Unity, installation of the S3/S4 arrays (left) give the ISS its definitive appearance for the first time. *(NASA)*

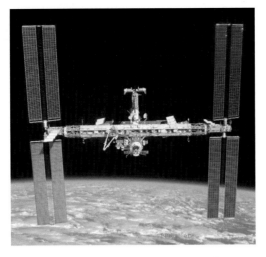

RIGHT The pre-assembled S3/S4 arrays mirror the P3/P4 assembly for location on the port extension of the truss segments. *(Boeing)*

BELOW In the Payload Canister that will deliver the S3/S4 segment to the Shuttle on the pad, engineers prepare the next element for flight. *(NASA)*

BELOW RIGHT Located at the extreme end of the S4 truss, the containers carry the support beams for the two opposing solar array wing pairs. *(NASA)*

with Williams and leaving Anderson with the Expedition 15 crew in her place.

S3/S4 Integrated Truss Segment

Length	44ft 9.6in (13.66m)
Width	16ft 3.4in (4.96m)
Height	15ft 2.3in (4.63m)
Weight	35,678lb (16,183kg)

The S3/S4 truss segments are the direct equivalent for the starboard side of the truss assembly as the P3/P4 segments were for the

port side, and incorporated the second Solar Array Alpha Rotary Joint (SARJ). When activated, they operated in conjunction with the SARJ in the P3/P4 arrays to turn the solar arrays through 360 degrees on each revolution of the Earth and maintain direct alignment with the sun for electrical power through the photovoltaic cells. References on the equipment carried by the S3/S4 truss assemblies are the same as those found in the P3/P4 payload description. S3/S4 was delivered to the Kennedy Space Center on 15 January 2001. Like the P3/P4 arrays, their delivery to the ISS was, of course, delayed by the *Columbia* disaster on 1 February 2003.

STS-118/ISS 13A.1/S5 Truss Segment

Endeavour
8 August 2007, 10.37pm GMT
Cdr: Scott J. Kelly (1)
Pilot: Charles O. Hobaugh (1)
MS1: Tracy E. Caldwell
MS2: Richard A. Mastracchio (1)
MS3: David R. Williams (Canada)
MS4: Barbara R. Morgan
MS5: B. Alvin Drew

E*ndeavour* was making its first visit back to the ISS since the STS-113 mission in November 2002 on NASA's 150th manned space flight launch. Following the loss of *Columbia* in 2003, *Endeavour* was extensively refurbished and modified for flight resumption on its 20th mission. Flying as a fully-fledged NASA mission specialist, Barbara Morgan had been back-up to Christa McAuliffe, winner of the Teacher in Space programme killed in the 1986 *Challenger* disaster.

The mission of ISS 13A.1 was to install the S5 truss segment and External Stowage Platform 3, and to restock the ISS with logistical supplies from the last Spacehab Logistics Single Module flown to the ISS. Set up by STS-116 the previous December, *Endeavour* had the advantage of the Station Shuttle Power Transfer System (SSPTS) which allowed the Orbiter to tap into the 120v DC power supply from the ISS, converting it to the 28v DC system on the Orbiter, relieving demand on the Shuttle's electricity-producing fuel cells and thereby affording an extra three days for emergencies.

LEFT MISSE, a material science experiments package, installed on the station exterior after transfer from *Endeavour*. *(NASA)*

Spacewalk time totalled 23hr 15min on four excursions outside the ISS. Mastracchio conducted three EVAs, Williams three, and Anderson two, completing installation of the S5 segment, cinching down the P6 radiator panels, changing out a failed Control Moment Gyro, relocating the CETA cart, and replacing external experiment packages. *Endeavour* undocked after a stay-time of 8 days 17hr 54min and landed back at the Kennedy Space Center on 21 August at an elapsed time of 12 days 17hr 55min.

ANATOMY

Spacehab Logistics Single Module
Previously carried aboard STS-116, the SLSM was described in the payload section for that mission. For STS-118 it carried cargo transfer bags, essential crew provisions including food, water, clothing, and personal items, and additional science equipment for conducting research aboard the ISS.

BELOW *Endeavour* **departs and gets an alien's-eye view of the ISS, the retracted P6 arrays appearing as a dislocated 'H'.** *(NASA)*

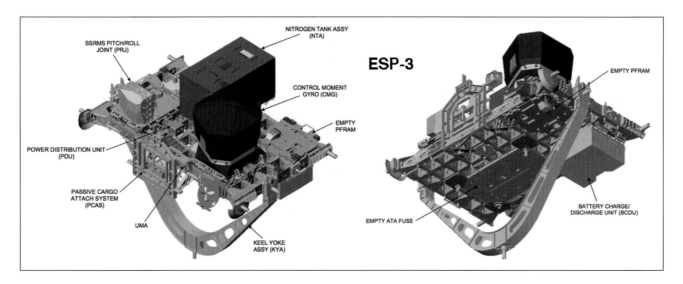

ESP-3

NITROGEN TANK ASSY (NTA)

SSRMS PITCH/ROLL JOINT (PRJ)

CONTROL MOMENT GYRO (CMG)

EMPTY PFRAM

POWER DISTRIBUTION UNIT (PDU)

PASSIVE CARGO ATTACH SYSTEM (PCAS)

UMA

KEEL YOKE ASSY (KYA)

EMPTY ATA FUSE

EMPTY PFRAM

BATTERY CHARGE/ DISCHARGE UNIT (BCDU)

ABOVE The third External Stowage Platform (ESP-3) carried to the ISS supported vital subsystems and replacement parts and would be moved around the exterior as required. *(NASA)*

BELOW Inside the Space Station Processing Facility at the Kennedy Space Center, engineers prepare the S5 spacer segment to be installed on S4. *(NASA)*

External Stowage Platform 3 (ESP-3)

Weight	7,495lb (3,400kg)

Similar in size to ESP-2, which was launched aboard STS-115, ESP-3 carried a group of ORUs on six of the seven attachment locations, including a Control Moment Gyroscope, a Nitrogen Tank Assembly, and a Pitch Roll Joint for Canadarm2. The carrier was lifted across to its location on the P3 truss segment, grasped by Canadarm2 and then moved down the truss assembly using the Mobile Transporter, the first such use of this robotic system to deploy a major payload. In January 2010 it was moved again to the lower part of the S3 segment, clearing the way for ExPRESS Logistics Carrier 3. Built by Spacehab, this was the last of three ESPs deployed to the ISS.

S5 Truss Segment

Weight	4,009lb (1,818kg)

Identical in almost every respect to the P5 truss used as a spacer between the P3/P4 and P6 truss segments, S5 served an identical function at the other end of the truss assembly on the starboard side and would reside there until the P6 array was relocated to this position from the Z1 truss. All information is found in the STS-116 payload description apart from the weight, stated above. S5 arrived at the Kennedy Space Center on 19 July 2001, 18 months before the Shuttle stand-down after the loss of *Columbia*.

Soyuz TMA-11/ISS S15/ Expedition 16

10 October 2007, 1.23pm GMT
Cdr: Yuri Malenchenko (Russia)
Flight Engineer: Peggy A. Whitson
Participant: Sheikh Muszaphar (Malaysia)

TMA-11 delivered the Expedition 16 crew along with short-stay visitor Sheikh Muszaphar, a Malaysian orthopaedic surgeon who took this flight via a special cooperative arrangement with Russia. TMA-11 docked at the ISS at 2.50pm GMT on 12 October, where for a few days six crewmembers conducted joint activities. Sheikh Muszaphar's mission lasted 10 days 21hr 14min from launch in TMA-11 and, landing in TMA-10, he returned to Earth on 21 October along with Expedition 15 crewmembers Kotov and Yurchikhin, leaving Anderson on board for a few days with the Expedition 16 crew until he returned aboard STS-120.

Expedition 16 was unusual in that the long-stay crew of Whitson and Malenchenko would continue for six months until April 2008, but host four NASA and ESA astronauts for various periods. Anderson returned aboard *Discovery* at the beginning of October and was replaced by Daniel Tani, who remained aboard the ISS until he in turn was relieved by ESA's Leopold Eyharts in February 2008 by *Atlantis*. Finally, Garrett Reisman relieved Eyharts in March 2008 via another *Atlantis* flight and he returned on *Discovery* in May 2008. In all, Expedition 16 would host three Shuttle ISS assembly visits, see the arrival of the Harmony node and ESA's first Automated Transfer Vehicle (Jules Verne) and host the first South Korean astronaut on a short-stay visit. The expedition also supported five spacewalks totalling 35hr 21min, at various times involving Malenchenko (1), Whitson (5), and Tani (4).

The first EVA, on 9 November, involved a lot of umbilical cables being connected on PMA-2 and Node 2, while the second, on 20 November, set up the external configurations of PMA-2 and the Harmony module. The third EVA was four days later, and continued outfitting PMA-2 and Harmony and provided further examination of the troublesome SARJ joint. The fourth EVA (on 18 December) allowed further inspection of the SARJ joint and of the Beta Gimbal Assembly, and was the 100th spacewalk in support of the ISS. The last EVA for this expedition, on 30 January 2008, was devoted to further study and testing of the SARJ and BGA assemblies. The resident Expedition 16 crewmembers Malenchenko and Whitson returned aboard Soyuz TM-11 on 19 April 2008 after a mission lasting 191 days, 19 hours and 17 minutes.

STS-120/ISS 10A/ Harmony (Node 2)

Discovery
23 October 2007, 3.38pm GMT
Cdr: Pamela Melroy (2)
Pilot: George D. Zamka
MS1: Douglas H. Wheelock
MS2: Stephanie Wilson (1)
MS3: Scott E. Parazynski (4)
MS4: Paolo A. Nespoli (ESA)
MS5 (up): Daniel M. Tani (1)
MS5 (down): Clayton Anderson

With a total payload of 40,872lb (18,539kg), STS-120 carried the Node 2 module installed on the port docking aperture on Node 1 (Unity). The P6 solar array from the Z1 truss was moved to the freshly installed P5 segment on the port side of the truss assembly. This completed installation of six of the eight

BELOW Scott Parazynski sets out to work repairing torn sections of the P6-4B solar array. *(NASA)*

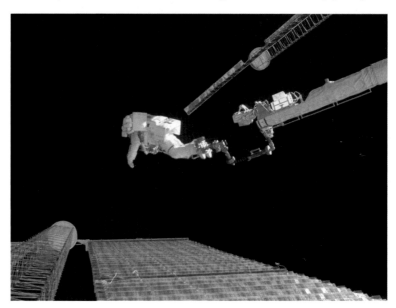

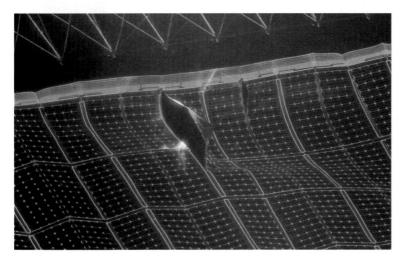

sets of paired Solar Array Wings. It took four EVAs totalling 27hr 14min conducted between 26 October and 3 November for Parazynski, Wheelock, and Tani to do the job, as well as examining once again the starboard SARJ joint, which was still proving troublesome. The fourth EVA enabled repairs to the damaged P6 4B solar array, torn in some places by extension after it had been relocated to P5. Before leaving, *Discovery* left Tani on the ISS to join Expedition 16, returning with Anderson, who had spent 152 days in space. After the Shuttle departed Canadarm2 was used to attach PMA-2 to the end of Node 2 and then relocate the combined assembly to the front of Destiny. Extended by one day to allow more time for the fourth EVA, *Discovery* logged 15 days 2hr 23min.

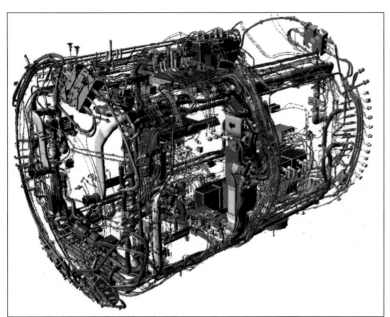

ANATOMY

Node 2: Harmony module

Length	23.6ft (7.2m)
Diameter	14.5ft (4.4m)
Pressurised volume	2,666ft^3 (75.5m^3)
Habitable volume	1,230ft^3 (34.8m^3)
Weight	31,500lb (14,288kg)

Harmony was the second of three node modules, interconnecting places where functional activities with adjacent pressurised modules were co-ordinated and where through-access could allow easy movement of people and cargo. Manufactured by Thales Alenia Space of Torino, Italy, under a contract to ESA, Node 2 incorporates six ports, two in the longitudinal axis and four mounted around the circumference of the Harmony cylinder at 90-degree intervals. Boeing provided several internal systems together with the Common Berthing Mechanism (see Node 1 Unity for description of various docking elements) and these were tested by Thales before delivery. Node 2 was named Harmony on 15 March 2007 following a competition involving 32,000 students in 32 US states. No fewer than six schools proposed the winning title, making it the first piece of ISS hardware not named by NASA.

LEFT Second of three Nodes, Harmony supported a network of fluid and gas pipes, electrical cables, data cables, communications equipment, life support systems and electronic subsystems. *(Thales Alenia)*

Harmony was launched with four avionics racks, two Resupply Stowage Racks fitted with cargo for the ISS, and two zero-g Stowage Racks. The node was initially attached to the port side of Unity (Node 1) directly opposite the Quest airlock, but when the PMA-2 adapter had been relocated from the front of Destiny to the end of Harmony, the new combination was moved to the front of Destiny, Harmony in effect being a spacer between PMA-2 and its original location on the front of the US laboratory. Harmony boosted the internal pressurised volume of the ISS's 15,000ft^3 (425m^3) by about 20 per cent, increasing it to 17,666ft^3 (500m^3). With the attachment of the Harmony module, NASA considered the ISS to be 'Core Complete', essentially having launched all the major pressurised US modules.

The agreement for Italy to build Nodes 2 and 3 was signed on 8 October 1997, and Harmony was flown from Europe to the US in an Airbus Beluga freighter and delivered to the Kennedy Space Center on 1 June 2003. Accompanying it on that flight was ESA astronaut Paolo A. Nespoli. Between the signing of the agreement with ESA and final fabrication, NASA cancelled a planned Habitation module and changed the function of Node 2 to incorporate living quarters for the US astronauts. Functionally, ESA's Columbus laboratory module would be attached to the starboard hatch and Japan's Kibo module would be attached to the port side, with the zenith and nadir ports housing logistics modules from the US and Japan as necessary.

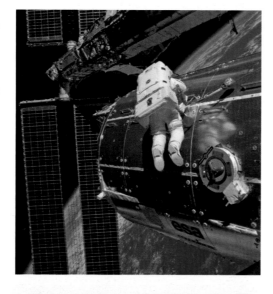

ABOVE Installed on the starboard side of the Harmony Node, Columbus is unpacked by ESA astronaut Hans Schegel for scientific experiments. *(NASA)*

RIGHT NASA astronaut Rex Walheim works outside Europe's Columbus module preparing it for operational use. *(NASA)*

BELOW Installed and ready for work, the ESA Columbus module attached to the starboard side of the Harmony module, also built by Europe. *(NASA)*

STS-122/ISS 1E/Columbus

Atlantis
7 February 2008, 7.46pm GMT
Cdr: Stephen N. Frick (1)
Pilot: Alan G. Poindexter
MS1: Leland D. Melvin
MS2: Rex J. Walheim (1)
MS3: Hans Schlegel (ESA) (1)
MS4: Stanley G. Love
MS5 (up): Leopold Eyharts (1)
MS5 (down): Daniel M. Tani (2)

Already delayed by two months due to a technical failure on the launch pad, after reaching orbit plans to deploy the Columbus module were delayed for a day due to a 'medical issue' with one of the crew. The first EVA, which should have been conducted by Rex Walheim and German astronaut Hans Schlegel, saw the ESA astronaut replaced by Stan Love. On 11 February the first EVA took place, during which the Columbus module was removed from the Shuttle's payload bay and the initial docking to the port side of Node 1 was made at 9.29pm GMT. European astronauts Schlegel and Eyharts entered Columbus for the first time at 2.08am GMT on 12 February, to much celebration in Europe and with congratulations offered by German Chancellor Angela Merkel. Thus did ESA become only the third space agency after Russia and the United States to fund, develop, build, and enter a pressurised space vehicle designed to remain in orbit.

A second EVA lasting 6hr 45min was conducted on 13 February to organise several adjustments to Columbus, installing power and data cables, and preparing to change out a nitrogen tank for the P1 truss. The third EVA took place two days later and lasted 7hr 25min, during which Walheim and Love installed the external experiment packages to Columbus and replaced a failed Control Moment Gyro removed on STS-118. *Atlantis* undocked at 9.24am GMT on 18 February and landed at an elapsed time of 8 days 16hr 7min, returning with Daniel Tani, whose Expedition 16 flight had lasted almost 132 days. Unfortunately, Tani's mission was marred by the news relayed to him on 19 December that his mother had been killed in an accident.

Columbus pressurised module

Length	22.5ft (6.9m)
Diameter	14.7ft (4.5m)
Weight (empty)	22,708lb (10,300kg)
Weight (launch)	26,627lb (12,077kg)
Max Payload Capability	19,845lb (9,000kg)
Max on-orbit mass	46,305lb (21,000kg)
Internal volume	2,648ft^3 (75m^3)
Volume of Payload Racks	883ft^3 (25m^3)

Designed and built as Europe's flagship presence aboard the ISS, Columbus has had a chequered history dating back to 1985 when the European Space Agency (ESA) decided to participate in the space station programme proposed by President Reagan in 1984 and sold to Canada, Europe, and Japan by the then NASA Administrator James M. Beggs. Europe committed to an Attached Pressurised Module (APM), a Man-Tended Free Flyer (MTFF) to provide European autonomy for independent space research, and an unmanned Polar Platform (PPF) designed to provide research opportunities in a polar orbit inclined approximately 90 degrees to the equator. The bulk of the development cost would be in the APM, which was envisaged as one of several international modules collectively forming the Freedom space station.

As development progressed ESA expanded its plans, proposing to build a mini-shuttle called Hermes launched by a powerful Ariane 5 launch vehicle developed primarily for lifting two or more satellites into geostationary transfer orbit. As projected costs escalated and budgets tightened, the polar platform was abandoned along with Hermes. To preserve work already conducted on respective projects, MBB-ERNO, the German prime contractor for the APM, negotiated a work share arrangement with Alenia in Italy for systems development and status as co-prime on detailed design and manufacturing. It was undoubtedly Italy's early entry to the difficult requirements of manned space flight that positioned that country's industry for building the Harmony and Tranquility nodes, Multi-Purpose Logistics Modules (MPLMs) and the Automated Transfer

Vehicle (ATV). Today, after a protracted series of name changes, MBB-ERNO is known as EADS Astrium Space Transportation.

Named after the Italian explorer from Genoa, Columbus owes much to the Spacelab pressurised modules utilised on Shuttle flights between 1983 and 1998 and consists of a pressurised cylindrical hull with closed welded end cones. The basic structure shares many design and manufacturing details with Europe's Multi-Purpose Logistics Modules and comprises primary and internal secondary structures built up from aluminium alloys. They are covered

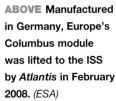

ABOVE Manufactured in Germany, Europe's Columbus module was lifted to the ISS by *Atlantis* in February 2008. *(ESA)*

LEFT Columbus is Europe's flagship module in the ISS partnership, a development of the Spacelab module carried aboard the Shuttle long before the ISS was planned. *(NASA)*

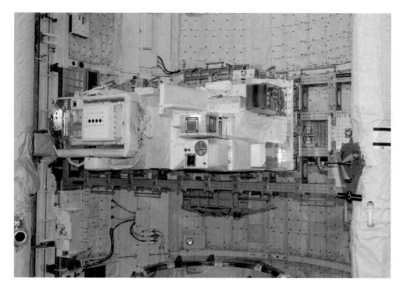

ABOVE Packed with equipment for transfer to the station, the Integrated Cargo Carrier was installed in *Atlantis*'s cargo bay. *(NASA)*

with more than 4,000lb (1,814kg) of aluminium panelling with layers of Kevlar and Nextel for micrometeoroid protection.

Columbus can accommodate 16 experiment or systems racks arranged around the circumference of the module in four sets of four racks at 90-degree intervals. All but six are of the International Standard Payload Rack (ISPR) configuration (see 'Experiments aboard the ISS') and because they are attached circumferentially, the truncated end cones house operating systems, cameras, switch panels, audio terminals, and fire extinguishers. The smallest of all the research modules

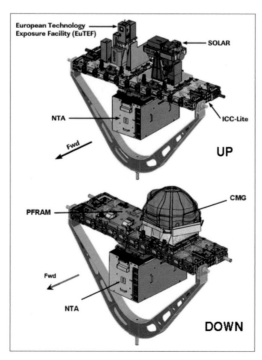

RIGHT The ICC carried European equipment up and was returned by *Atlantis* with a failed control moment gyroscope (CMG). *(NASA)*

constituting the ISS, the design is such that it contains exactly the same payload volume, power, and data facilities as the others with 5,511lb (2,500kg) of experiment equipment preloaded at launch.

The Biolab carries equipment for studying micro-organisms, cell and tissue cultures and can accommodate small plants and animals. The Fluid Science Laboratory is dedicated to the study of fluids in the micro-g environment for better understanding of energy production, propulsion, and environmentally improved ways of managing resources. The European Physiology Modules Facility is dedicated to the study and evaluation of the human body. Functions such as circulation, respiration, bone loss, and the response of the immune system to weightlessness not only helps physicians understand how best to prepare humans for very long flights, but the research also feeds directly back into the lives of people on Earth. Many medical benefits and improved understanding of how the body ages stems from this research, which is being integrated into the general understanding of what happens to the body over time as populations get older.

All these research facilities are designed for autonomous operation as fully as practicable, freeing astronauts to conduct hands-on experiments and general station 'housekeeping' duties. Externally, Columbus has four mounting points for payloads and packages dedicated to Earth observation, space technology, and innovative new space-based processes. One such pre-loaded attachment was the European Technology Exposure Facility (EuTEF), carrying a range of devices which are exposed to space, weighing only 771lb (350kg) and requiring only 450W of power. A second attachment was the SOLAR observatory, which conducted a spectral survey of the sun for the first two years after launch. Later, additional external experiments were added, first of which was the Atomic Clock Ensemble in Space (ACES) device for testing the next generation of cold-atomic clocks, and later the Atmosphere Interaction Monitor which studies the coupling of thunderstorms and processes in the upper atmosphere, the ionosphere, and the Van Allen radiation belts.

Three rack positions were dedicated to subsystems essential to running the module –

items such as water pumps, heat exchangers, and avionics packs. Columbus is capable of receiving up to 20kW of electrical power, of which 13.5kW can be used for experiments and research equipment. To maintain an acceptable environment for up to three people and for ventilation, Columbus has a continuous airflow from Node 2 to which it is attached. Recirculated air flows back to Node 2 where the carbon dioxide is scrubbed out, the temperature inside the laboratory being controlled to within 61–86°F (16–30°C) with varying humidity levels. Columbus has its own water loop for connecting to the ISS heat removal system and an air/water heat exchanger removes excess condensation. Electrical heaters can supplement atmospheric recirculation when the attitude of the ISS places Columbus in a cold-soak out of direct sunlight.

Command of the module takes place at the Columbus Control Center at Oberpfaffenhofen in Germany, located on the premises of the German Space Operations Centre (Deutches Zentrum für Luft- und Raumfahrt, or DLR). European payloads aboard Columbus send their data direct to this facility, where all experiments are co-ordinated and managed, and all activities are synthesised with NASA's Mission Control Center in Houston.

ATV1 Jules Verne/ ISS-ATV1

9 March 2008, 4.03am GMT

Another triumph for European engineering design and technology, the Automated Transfer Vehicle was launched by Ariane 5 from the ELA-3 launch complex at ESA's facility at Kourou in French Guiana on the north-east coast of South America. After achieving orbit the Vulcain engine of the second stage fired twice to adjust the orbit, ATV1 separating from the stage at 1hr 7min into the flight at which point the module was switched on for a series of housekeeping checks and for setting up the propulsion system for rendezvous and docking manoeuvres. This necessitated a firing of the four main engines on 11 March. Because this was the first flight of the ATV, three weeks of orbital testing preceded docking.

LEFT Launched by Europe's Ariane V from French Guiana, along with the Russian Progress cargo tanker the ATV is the only logistics module designed to rendezvous and dock with the ISS, those built by Japan and by commercial companies in the US being grappled by Canadarm2 and manually hooked up to the station. *(ESA)*

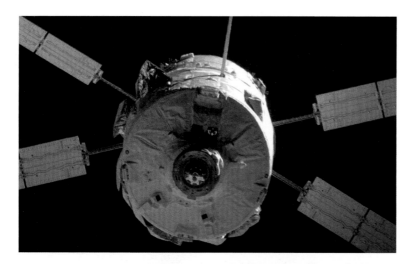

ABOVE The forward docking port on ATV-1 as it slowly approaches the aft docking port on Russia's Zvezda module. *(ESA)*

RIGHT With copies of Jules Verne's work, crewmembers enjoy the spacious and noticeably quiet interior of ATV-1. *(ESA)*

BELOW In this artist's depiction of the ISS over the Netherlands, the ATV is attached to the aft end of Zvezda, other modules in line to the right being Zarya, Unity and Destiny. Noticeably absent is ESA's Columbus but the Quest airlock is visible on the side of Unity below the P6 truss. *(ESA)*

First the solar arrays were deployed and at 10 days into the flight it was positioned about 1,080nm from the ISS where it would maintain formation at what was known as the S1 position awaiting the completion of the STS-123 mission in late March. Under command from the ATV control centre, the module was gradually moved up to its target for a series of demonstrations to prove that in the event of a problem close to the station a collision-avoidance manoeuvre could be performed, these being conducted on 13–14 March. The GPS navigation systems were successfully tested as ATV1 moved to the S2 position approximately 1.9nm away from the ISS, where an escape manoeuvre was demonstrated on 29 March.

A second demonstration test on 31 March took ATV1 to within 35ft (10.7m) of the aft docking port on Zvezda at the S3 position, followed by a simulated escape move taking it 65ft (19.8m) away from the ISS to S4. Only then, when all the systems, subsystems, and operating routines had been tested, demonstrated, and qualified could it make the final approach and dock to the Zvezda module. With authorisation from Moscow mission control, permission was granted for the final approach. Using the Russian Kurs radar system, GPS, and onboard laser instruments for ranging and approach-rate measurements beginning at a distance of just over 800ft (244m), ATV1 hooked up at 2.45pm GMT on 3 April. After leak checks and pressure equalisation the hatch was opened and the transfer of cargo could begin, the capacious volume thus created being used to stash waste and used items for burn-up in the Earth's atmosphere.

Among items carried to the Expedition 16 crew aboard the ISS were two original manuscripts by Jules Verne and a French 19th-century edition of *From the Earth to the Moon* and *Around the Moon*. On 27 August the ATV1 propulsion system was used to slow the docked assembly by 2.2mph (3.5kmh) to avoid possible collision with a piece of debris from the Cosmos 2421 satellite, lowering altitude by 1.1nm. Items transferred to Zvezda included 600lb (272kg) of water, 46lb (21kg) of oxygen, and 1,890lb (857kg) of propellant, as well as 2,500lb (1,134kg) of dry cargo for the

ISS in general. The ATV was a noticeably quiet place and was used by the ISS crew unofficially as a place to sleep and to perform personal hygiene, while the South Korean astronaut Yeo So-Yeon also used ATV1 for conducting some nanotechnology experiments.

The module was sealed up and it undocked at 9.29pm GMT on 5 September, from where, after moving away about 3 miles below the ISS, its first de-orbit manoeuvre took place. Starting at 10.00am GMT on 5 September with a 6-minute firing of its main engines, a second firing lasting 15 minutes followed at 12.58pm GMT. ATV1 encountered Earth's atmosphere at 1.31pm GMT and disintegrated within 12 minutes during its fiery re-entry.

Anatomy

ATV1 Jules Verne

Length	32.1ft (9.8m)
Diameter	14.7ft (4.5m)
Solar array span	73.1ft (22.3m)
Dry Weight	23,086lb (10,472kg)
Consumables	5,762lb (2,614kg)
Cargo capacity	16,537lb (7,500kg)
Weight at launch	45,753lb (20,753kg)
Waste capacity for download	13,890lb (6,300kg)

The Automated Transfer Vehicle was designed as a logistics and resupply cargo module for uploading liquid and solid freight to the ISS, returning to destructive re-entry with more than six tonnes of waste. Built by European companies as part of ESA's contribution to the ISS, ATVs are launched by Ariane 5 from the Arianespace launch site in French Guiana. The main pressure shells are fabricated by Alenia Spazio in Turin, Italy, and the propulsion sections are built by EADS Astrium in Bremen, Germany. Docking and refuelling systems for connecting to the Zvezda module are supplied by RSC Energia in Russia and the integrated system was tested at the ESTEC facility at Noordwijk in Holland. Because they can be used only once, ESA funds a modest ATV production line with flights originally planned for every 18 months but that has been adjusted, with almost three years between ATV1 and ATV2 and flights scheduled annually thereafter.

© ESA/D.Ducros - 2010

ABOVE Europe's first Automated Transfer Vehicle (ATV-1) was named after the science fiction writer Jules Verne and comprised an unpressurised service module (aft) and a pressurised cargo section (forward). *(ESA)*

BELOW The ATV can carry more than 10 tons of dry and wet cargo with ready access, doubling as a refuse truck which, along with its contents, burns up on re-entry. *(ESA)*

The ATV consists of two primary sections: the pressurised Integrated Cargo Carrier (ICC) and the unpressurised Service Module. The ICC is a cylindrical structure manufactured from 2219 aluminium with cone end-sections, the forward section supporting the docking probe and hatch, the aft section mated to the Propulsion Module. It is 16.1ft (4.9m) in length and 14.7ft (4.48m) in diameter and is protected externally by a micrometeoroid and orbital debris protection system consisting of Nextel/Kevlar blankets, with a thermal insulation of gold-coated Kapton insulation blankets and aluminised beta cloth. The Service Module supports attachment and deployment mechanisms for the four solar arrays, contains water and gas tanks at the forward end and the four main rocket motors for orbital manoeuvring on the aft face. It has a length of 11.2ft (3.4m) including the shaped adapter increasing its diameter to that of the ICC.

The ATV is highly autonomous, with its own propulsion and station re-boost system, avionics equipment, guidance navigation and control systems, communications capability, electrical power generation and storage, and rendezvous and docking systems. It conducts the same role as the Russian Progress cargo-tanker vehicles but it has twice the carrying capacity and much greater internal volume. It can carry up to 12,127lb (5,500kg) of dry goods, 220lb (100kg) of oxygen and nitrogen, up to 1,896lb (860kg) of propellants for Zvezda or up to 10,363lb (4,700kg) of attitude control and re-boost propellant. An environmental control system provides air circulation, temperature monitoring, and fire detection sensors. Thermal control is both passive, with thermal insulation blankets, and active, with variable and constant conductive coolant pipes and paint surfaces. The ATV can support eight standard racks in pairs on all four interior surfaces.

The four Solar Array Wings are arranged in pairs mounted at 44 degrees to each other on opposing sides of the module. They are hinged in four segments each, restrained against the side of the Service Module for launch and only deployed after separating from the second stage of the launch vehicle. The power production and storage system has 40amp/hr rechargeable batteries and delivers 3.8kW of

which 400W will maintain the ATV in dormant mode with 900W drawn down for active operation. The main propulsion system consists of four liquid bi-propellant rocket motors placed at 90-degree intervals around the aft-facing end section of the Propulsion Module, each with a thrust of 110lb (50kg) consuming MMH (monomethyl hydrazine) and N_2O_4 (nitrogen tetroxide). Attitude control is achieved by 28 liquid motors using the same propellant combination as the main engines, each with a thrust of 49.5lb (22.45kg).

The first ATV, Jules Verne, was named after the 19th-century French author and visionary. The second, ATV2 Johannes Kepler, was named after the German astronomer and mathematician who defined the elliptical orbits prescribed by all planetary bodies, and was launched on 16 February 2011. The third, ATV3 Edoardo Amaldi (which was launched on 23 March 2012), was named after the leading Italian scientist of the 20th century, one of a select few under Enrico Fermi who discovered the existence of slow neutrons. Scheduled for launch in February 2013, ATV4 Albert Einstein has been named after the physicist who laid down fundamental laws of relativity uniting space and time in theoretical physics. ATV5 has yet to be named and will be launched in February 2014.

STS-123/ISS 1JA/Kibo Experiment Logistics Module

Endeavour
11 March 2008, 6.28am GMT
Cdr: Dominic L. Gorie (3)
Pilot: Gregory H. Johnson
MS1: Robert L. Behnken
MS2: Michael Foreman
MS3: Takao Doi (1) (Japan)
MS4: Richard M. Linnehan (3)
MS5 (up): Garrett E. Reisman
MS5 (down): Leopold Eyharts (ESA)

A routine mission but a significant one in that it delivered Japan's Kibo Experiment Logistics Module-Pressurised Section (ELM-PS), the first of Japan's major contributions to

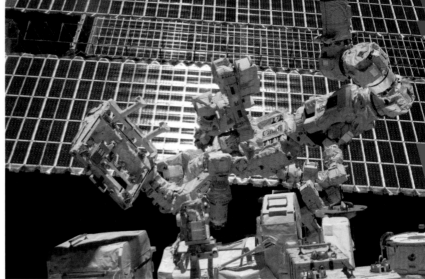

ABOVE **Dextre is lifted by Canadarm2 with infinite flexibility to operate alone or in conjunction with the other manipulator at the ISS.** *(NASA)*

ABOVE **Dextre and Canadarm2 conducting remotely operated chores around the exterior of the space station.** *(NASA)*

LEFT **Japan's first contribution to the ISS, the Kibo Experiment Logistics Module was carried to the station by *Endeavour* and stowed temporarily on Harmony.** *(NASA)*

the ISS. Successive Shuttle flights would deliver the pressurised module and an unpressurised experiments platform. Japan had been an early convert to the concept of an international station and made a greater contribution than Europe in achieving a larger percentage of the overall experiment space than other countries. *Endeavour* also carried Dextre, the Canadian Special Purpose Dextrous Manipulator (SPDM).

Endeavour docked with the PMA-2 port at 4.49am GMT on 13 March and Dextre was moved across on its pallet to the POA attachment on the Mobile Base Station, completing the suite of Canadian robotic equipment delivered to the ISS. But the pallet could not get power to allow autonomous operation and the Shuttle's RMS had to be used to lift the Kibo ELM from the payload bay and install it at its interim position on the Harmony Node 2 zenith (upper) port at 8.06am GMT on 15 March. In five spacewalks lasting 33hr 28min Linnehan, Reisman, Foreman, and Behnken installed the ELM-PS and the SPDM robot, conducted several maintenance tasks, replaced a Power Control Module and retrieved experiment packages.

Because there would be insufficient room in the cargo bay of STS-124 due to the size of the Kibo Japanese pressurised module, the Shuttle's Orbiter Boom Sensor System was left at the ISS when *Endeavour* returned, to be retrieved and brought back by STS-124 two months later. *Endeavour* made use of the

Station to Shuttle Power Transfer System to extend the flight's duration to a record 11 days 20hr 36min docked to the ISS, departing on 26 March with ESA's Leopold Eyharts, who had been a part of the Expedition 16 crew since arriving in February, and leaving Garrett Reisman to bridge that crew and Expedition 17. Total mission duration was 15 days 18hr 11min.

Anatomy

Kibo Experiment Logistics Module-Pressurised Section (ELM-PS)

Length	13.8ft (4.2m)
Outer diameter	14.4ft (4.4m)
Inner diameter	13.8ft (4.2m)
Weight	18,490lb (8,387kg)
Environment	65–86°F (18–30°C)
Humidity	25–70%
Power	3kW at 120VDC

First to be delivered of three Kibo elements, the ELM-PS provides storage facilities for experiment payloads, for samples, and for spare parts necessary in other Japanese elements. Free access between the pressurised modules is afforded in orbit and is the only onboard facility with its own storage provision. Packed with supplies on the way up, the ELM-PS would be used in orbit for stowing equipment and experiments utilised in the main Pressurised Laboratory Module (PLM) and on the external pallet. The ELM-PS was attached to the upper (zenith) port on Harmony until the arrival of the Pressurised Laboratory Module on STS-124, when it was moved across from the top of Harmony to the zenith port of the Kibo PLM.

Essentially a cylindrical pressure module with truncated end cones, the ELM-PS is a structural body following the same basic design rationales of other pressure modules attached to the ISS. Designed to accommodate launch and ascent loads and to have a design life of ten years, the module is fabricated from aluminium alloy panels covered with external debris shields. It has a Flight Releasable Grapple Fixture (FRGF) for robotic arms to grasp it for relocation, a Common Berthing Mechanism (CBM) for

berthing with the PLM, and an Exposed Facility (EF) from which payloads or cargo can be unloaded when the H-II Transfer Vehicle (HTV) brings supplies and equipment.

The Kibo ELM-PS was shipped from Japan to the Kennedy Space Center in April 2003, arriving two months later. During launch and until unpacked, the ELM-PS carried eight modules. Five carried electrical power, data management and communication subsystems, and two were work stations. Two carried experiment racks and one was a stowage rack. Some of the subsystem racks would be transferred to the PLM after that had been attached to Harmony.

Special Purpose Dextrous Manipulator (SPDM)

Height	12ft (3.7m)
Width across shoulders	7.7ft (2.3m)
Arm length (each)	11ft (3.4m)
Weight	3,440lb (1,560kg)
Handling capacity	1,323lb (600kg)
Positioning accuracy (incremental)	2mm
Positioning accuracy (relative to target)	6mm
Force accuracy	0.5lb (227g)
Average operating power	1.4kW

Otherwise known as Dextre, the SPDM is the third and final element in Canada's robotic contribution to the ISS, following Canadarm2 and the Mobile Base System. In appearance it resembles the human torso, with arms and manipulators replicating the physical features of the upper body. Designed to move large loads, perform delicate manipulation of small objects, and be controlled by operators inside the ISS or on the ground, it has lights, video cameras, a stowage platform, and three common robotic tools.

Just as Canadarm2 was a development of the Shuttle's Remote Manipulator System, so too is Dextre a further development of the latter, allowing it to grasp ISS components with one hand while holding itself firm with another, opening and closing apertures, removing or replacing covers, or deploying or retrieving equipment. For grasping it has grips with a

BELOW Dubbed Dextre, or SPDM (Special Purpose Dextrous Manipulator), this technically advanced manipulator was the third of Canada's robotic contributions after Canadarm2 and the Mobile Transporter. *(MDRobotics)*

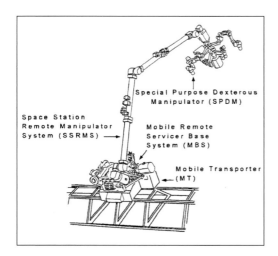

ABOVE The relative integration of the station robotic systems allows for motion above the 'rail track' that forms the full length of the completed truss assembly. (NASA)

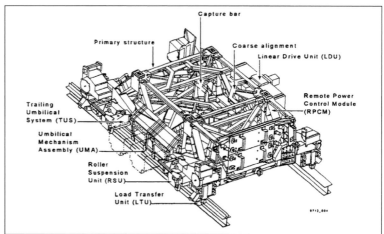

ABOVE The Mobile Transporter gives mobility to everything it carries, including astronauts. (NASA)

BELOW The Base System sits on the Mobile Transporter and is a location station for the manipulators or astronauts. (NASA)

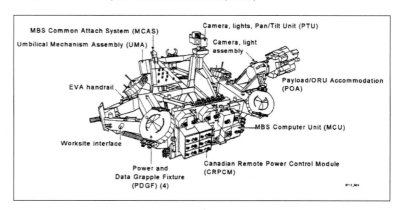

built-in socket wrench, camera, and lights, and two pan-and-tilt cameras below its rotating torso provide multi-angle views of its work site. Dextre is capable of conducting full engineering maintenance duties including the complete replacement of systems units, bolting and unbolting equipment, and working at millimetre-level accuracy for aligning and inserting tools and for tightening and untightening bolts.

A major advance in the design of Dextre was to give it sensitivity measuring forces and

BELOW Dextre is the most advanced robotic system installed on the exterior of the station and is the crew's exterior pair of hands for moving equipment around and playing a role in replacing equipment with the need for a space walk. (NASA)

BELOW Prior to launch, laid out ready for packing aboard *Endeavour*, Dextre is given scale by the technicians. (NASA)

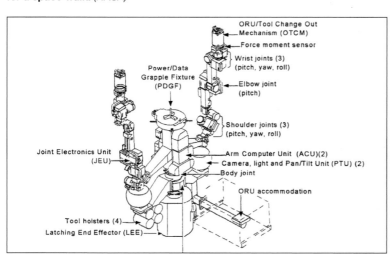

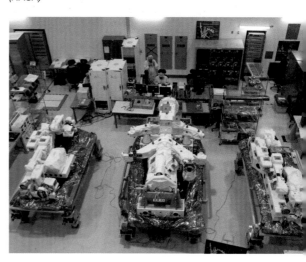

torque with an auto-compensation capability to minimise the possibility of damage to extremely delicate structures and devices. Dextre can pivot at the waist and its two shoulders each support an arm with seven offset joints to allow total freedom of movement. This allows the waist to change the overall position of the arm orientation without moving them independently and also restricts movement to only one arm at a time.

Dextre can maintain its grip while operating in conjunction with the Shuttle's RMS (until the Shuttle was retired) and with Canadarm2 on the ISS. A single device at the end of each arm doubles as a tool-changing mechanism incorporating a retractable motorised socket for turning bolts or detaching items such as cameras and lights and a retractable cable feeds power, data, and video connectors to payloads. In this way, through the interconnections, Canadarm2, the Mobile Base, and Dextre can be controlled simultaneously in a coordinated way.

Soyuz TMA-12/ISS 16S/ Expedition 17

8 April 2008, 11.17pm GMT
Cdr: Sergei Volkov (Russia)
Flight Engineer: Oleg Kononenko (Russia)
Participant: Yi So-Yeon (South Korea)

BELOW Permanently attached to the port side of the Harmony Node, the Kibo laboratory supported the Pressurised Logistics Experiment Module on top. *(JAXA)*

Carrying an astronaut from the South Korean space programme along with two Russian cosmonauts, TMA-12 docked with the Pirs module at 11.16am GMT. Yi So-Yeon returned in TMA-11 along with the Expedition 15 crew

after 10 days and 21 hours on her first and only space flight, leaving Expedition 16 to continue. Expedition 17 was in two parts, the first until June when Garrett Reisman was hosted aboard the station, succeeded by Gregory Chamitoff, who arrived with the docking of STS-124 on 2 June. Volkov and Kononenko returned aboard TMA-12 on 24 October after a mission duration of 199 days 4hr 20min, taking with them space tourist Richard Garriott, whose mission had lasted 11 days 20hr 36min. Chamitoff remained on the ISS to host the Expedition 18 crew until he too returned in STS-126 in November.

STS-124/ISS1J/Kibo Pressurised Laboratory Module

Discovery
31 May 2008, 9.02pm GMT
Cdr: Mark E. Kelly (2)
Pilot: Kenneth T. Ham
MS1: Karen L. Nyberg
MS2: Ronald J. Garan Jr.
MS3: Michael E. Fossum (1)
MS4: Akihiko Hoshide (Japan)
MS5 (up): Gregory E. Chamitoff
MS5 (down): Garrett E. Reisman

Discovery carried the Kibo module and docked it to the port side of the Harmony module during the first EVA on 3 June. It was exactly 43 years after the first US spacewalk conducted by Ed White from *Gemini III* in 1965. White died in the *Apollo* launch pad fire of 27 January 1967, but this anniversary was remembered as Fossum and Garan began their 6hr 48min EVA. At 9.05pm GMT the following day the hatch to Kibo was opened. The second EVA conducted on 5 June lasted 7hr 11min, during which Fossum and Garan installed covers and external TV equipment on the PLM and conducted sundry tasks.

On 6 June the Kibo ELM-PS lifted to the ISS by STS-123 was relocated from the zenith port on Harmony to the zenith port on the Kibo PLM, and the following day Hoshide and Nyberg, working in the Kibo module, unlatched the module's robotic arm and began tests. The third and final EVA on 8 June lasted 6hr 33min, during which Garan rode the robotic arm extension with the OBSS some 88ft (26.8m) above the ISS to replace a 550lb (250kg) nitrogen tank on the starboard truss.

Large quantities of stores and logistical supplies were offloaded to the ISS, 34,353lb (15,582kg) including the PLM of which 1,787lb (811kg) was to the interior of the station, with 1,807lb (820kg) transferred from inside the ISS to *Discovery*. Chamitoff replaced Reisman aboard the ISS and *Discovery* separated from the ISS after a docked duration of 8 days 17hr 39min. Touchdown at the Kennedy Space Center occurred at 13 days 18hr 13sec elapsed time on 14 June.

Anatomy

Kibo Pressurised Laboratory Module (PLM)

Length	36.7ft (11.2m)
Diameter	14.4ft (4.4m)
Weight	32,600lb (14,787kg)

Main Arm Remote Manipulator System	
Length	33ft (10m)
Weight	1,719lb (780kg)
Handling capacity	3,175lb (1,440kg)
Positioning accuracy translation	+/– 2in (5cm)
Positioning accuracy rotation	+/– 1 degree
Translation speed 1,323–6,614lb	2.4in (61mm)/sec
Translation speed greater than 6,614lb	1.2in (30mm)/sec

Fine Arm Remote Manipulator System	
Length	7.2ft (2.2m)
Weight	419lb (190kg)
Handling capacity with compliance mode	176lb (80kg)
Handling capacity without compliance mode	661lb (300kg)
Position accuracy translation	0.45in (11.4mm)
Position accuracy rotation	+/–1 degree
Translation speed less than 176lb	2.4in (61mm)/sec
Translation speed 176–661lb	1.2in (30mm)/sec

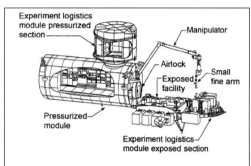

The PLM comprises the second of Japan's three Shuttle payloads to the ISS, the first having been uplifted by STS-123 and the last comprising the Kibo Exposed Facility (EF) and the Experiment Logistics Module-Exposed Section (ELM-ES) to be carried aboard STS-127 in July 2009. The three elements were designed into Japan's long-term science, technology, and research programme as an integrated part of the capabilities enhanced by the H-II Transfer Vehicle (HTV), a logistics carrier similar in function to Europe's Automated Transfer Vehicle (ATV) and launched by Japan's H-II rocket from Tanegashima, Japan.

The PLM is the largest by physical size of all

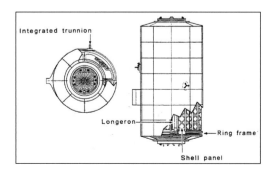

ABOVE Kibo incorporated its own robotic systems at the exposed end of the Pressurised Laboratory Module. *(JAXA)*

LEFT Japan's commitment to the ISS involved three separate modular elements: the Pressurised Laboratory Module, the Pressurised Logistics Experiment Module and the Exposed Experiment Logistics Module. *(JAXA)*

LEFT Similar in engineering design to the US Destiny and ESA Columbus modules, the Pressurised Laboratory Module was of waffle construction for rigidity, light weight and stiffness, defined by longerons and ring frames. *(JAXA)*

the pressurised modules attached to the ISS, containing up to 23 racks of which ten are the International Standard Payload Rack (ISPR) size (see Experiments Section). Two grapple fixtures are attached to the exterior to which a robotic arm can be attached for moving the PLM. It has a small scientific airlock through which instruments can be exposed to the space environment. Items passed through the cylindrical airlock must be smaller than 1.5ft x 2.7ft x 2.6ft (46cm x 82cm x 79cm) and limited to 661lb (300kg) in mass by the capacity of the slide table conveying the items into the airlock. The inner hatch is manually opened and closed while the outer door has a motorised hatch that retracts inwards. Equipment passed through the hatch is grasped by a small fine-sensor arm on a remote manipulator attached to the outside of the module, visually observed via a small window in the airlock door.

The PLM also has two windows and a unique berthing mechanism for attaching the EF to the PLM. In size, the PLM has the equivalent length of an eight-rack space but the airlock and the CBM unit limits rack installation. Six racks can be attached in a row to three of the four internal surfaces, or walls, with a five-rack row on the zenith surface, providing 23 rack spaces in all. On the exterior, an Exposed Facility Berthing Mechanism (EFBM) is used to connect the Exposed Facility to the PLM, EFBM-A on the Pressurised Module and EFBM-B on the EF.

The robotic arm attached to the exposed end of the PLM is known as the Japanese Experiment Module Remote Manipulator

BELOW The Kibo Pressurised Laboratory Module was lifted to the ISS by *Discovery* in May 2008.

System (JEMRMS) and is approximately two-thirds the size of the Shuttle RMS. In fact it consists of two separate arms, a 33ft (10m) long main arm and a 6ft (1.8m) long fine arm, each equipped with six independent joints. Both arms carry remotely operated cameras and from the inside of the PLM both arms can be controlled independently of each other. Being of modular design, components can be exchanged or replaced with some being replaceable from the interior.

Soyuz TMA-13/ISS 17S Expedition 18

12 October 2008, 7.01am GMT
Cdr: Yuri Lonchakov (Russia)
Flight Engineer: Michael Fincke (US)
Tourist: Richard Garriott (US)

A routine expeditionary changeover mission and Soyuz rescue vehicle replacement flight, TMA-13 also carried video-gaming entrepreneur Richard Garriott, son of NASA astronaut Owen Garriott, a veteran of the third manned Skylab flight launched in 1973 and of the first Spacelab mission in 1983. Richard Garriott returned with the Expeditionary 17 crew aboard TMA-12 on 24 October after a duration of 11 days 20hr 36min. Expedition 18 would encompass the resident long-stay crew of Fincke (commander of the expedition but not of the Soyuz flight, which is always a Russian) and Lonchakov with Chamitoff bridging the end of the preceding expeditionary crew and the arrival of Sandra Magnus aboard STS-126 in November. She in turn would be replaced by Koichi Wakata for the last month of the expeditionary mission.

The expedition supported two spacewalks. The first, on 23 December, lasted 5hr 38min while Lonchakov and Fincke installed a probe on the Pirs module for measuring electromagnetic energy and removed and replaced various external science packages. The second EVA, on 10 March 2009, was also conducted from the Pirs module, with the two spacewalkers replacing science packages and conducting a photo-survey of the Russian modules. The Expedition 18 mission ended on 8 April 2009 at 7.16am GMT with a duration of 178 days and 15 minutes.

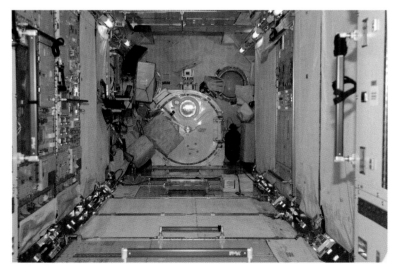

STS-126/ISS ULF2

Endeavour
15 November 2008, 12.56am GMT

Cdr: Chris Ferguson (1)
Pilot: Eric Boe
MS1 (up): Donald Pettit (1)
MS2: Steve Bowen
MS3: Heidemarie Stefanyshyn-Piper (1)
MS4: Shane Kimbrough
MS5 (up): Sandra Magnus (1)
MS5 (down): Greg Chamitoff

Endeavour flew a standard resupply mission paired with an urgent need to repair the starboard SARJ joint that had plagued operations with the ISS since first observed in August 2007. The Leonardo MPLM was packed with more than 14,000lb (6,350kg) of supplies including two additional crew racks comprising a second galley and a second waste and hygiene compartment. Four spacewalks totalling 26hr 41min conducted on alternate days starting on 18 November saw intense efforts at lubricating and servicing the starboard SARJ bearings and relocation of the two CETA carts along the truss. Leaving Pettit at the ISS and returning with Chamitoff, Endeavour spent 11 days 16hr 48min docked to the ISS and landed at an elapsed time of 15 days 20hr 29min.

STS-119/ISS 15A/S6 Truss Segment

Discovery
15 March 2009, 11.44pm GMT

Cdr: Lee Archambault (1)
Pilot: Tony Antonelli
MS1: Joseph Acaba
MS2: Steve Swanson (1)
MS3: Richard Arnold
MS4: John Phillips (1)
MS5 (up): Koichi Wakata (Japan) (2)
MS5 (down): Sandra Magnus (1)

This was the 100th Shuttle flight since the loss of Challenger on 28 January 1986, and once again, as they had been for every flight since STS-26, a bunch of roses arrived at Mission Control from the Shelton family of

Bedford, Texas. This time, delivered personally. Due to configuration alignments the S6 truss assembly could not be manoeuvred out of the Shuttle's cargo bay by the RMS so on 18 March Canadarm2 got that job, reaching in to retrieve it before handing S6 over to the RMS. While the RMS moved S6 over to a different location, Canadarm2 was moved along the truss assembly by the Mobile Transporter, receiving S6 back for a parked position overnight.

Installation came the following day during the first EVA. This spacewalk lasted 6hr 7min,

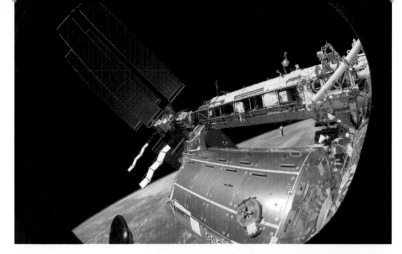

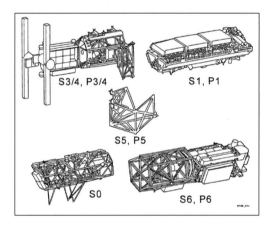

in which Swanson and Arnold monitored the installation of S6 into the S5 segment and connected cables and electrical conduits. They removed launch restraints, keel pins, and thermal covers, and deployed the S6 PV Radiator. The next EVA two days later had Swanson and Acaba prepare the exterior of the ISS for the STS-127 mission and the final Kibo elements in a spacewalk lasting 6hr 30min.

Two days after that, Acaba and Arnold spent 6hr 27min relocating various items of equipment and attempting to deploy a cargo carrier attachment system but failed, a repair job on that device being successfully conducted during the STS-127 mission. On 25 March the station and Shuttle crews talked by telephone to President Obama and members of Congress and closed the hatches before undocking after 7 days 22hr 33min at the ISS. Leaving astronaut Wakata at the station, Magnus returned with *Discovery* touching down at an elapsed time of 12 days 19hr 29min. Magnus's space flight had lasted for 21 minutes short of 130 days.

RIGHT Common elements for port and starboard truss segments minimise the variety of different designs and reduces manufacturing costs. *(Boeing)*

BELOW The Solar Array Wing design is common to the outboard truss segments. *(Boeing)*

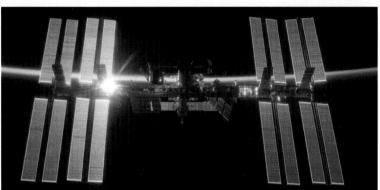

Anatomy

S6 Truss Segment

Length	45.4ft (13.8m)
Width	16.3ft (5m)
Height	14.7ft (4.5m)
Weight	31,060lb (14,089kg)

The payload for this mission comprised the last truss segment installed to the ISS, the equivalent on the starboard side of the configuration to the port (P6) array that had been providing power for the station since it was attached in November 2000. Adding the final pair of Solar Array Wings and associated radiator panels, S6 would further boost power and clear the way for a permanent six-person expeditionary crew.

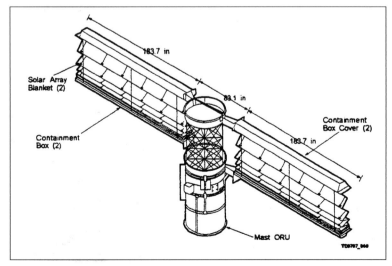

Unique to S6, the array also carried two spare Battery Charge/Discharge Units (BCDUs) for controlling the spare batteries on the ISS.

With S6, the 310ft (94m) long truss assembly was complete and formed the backbone for the ISS. Built as an operational test unit for its sister truss segment, P6, the S6 element was delivered to the Kennedy Space Center on 17 December 2002, its launch held in a stack of delivery and assembly flights grounded after the loss of *Columbia* in February 2003. Built primarily by Boeing, it was equipped with electrical equipment mostly provided by Pratt & Whitney Rocketdyne. All other pertinent details can be found on the payload description for the P6 truss assembly.

The addition of S6 added 66kW (21–30kW usable of which 15kW for science) bringing the total ISS generation capacity to 264kW, of which 84–120kW was of usable power with 30kW for scientific experiments and equipment.

Soyuz TMA-14/ISS 18S Expedition 19

26 March 2009, 11.49am GMT
Cdr: Gennady Padalka (Russia)
Flight Engineer: Michael Barratt (US)
Tourist: Charles Simonyi

TMA-14 delivered the Expedition 19 crew and space tourist Charles Simonyi, who was making his second self-funded flight – his first being aboard TMA-10 in April 2007. Simonyi returned with the Expeditionary 18 crew on 8 April at 7.16am GMT, a mission duration of 12 days 19hr 27min. This was the final three-person expeditionary crew before the start of a full six-person complement aboard the ISS, Padalka and Barratt joining Koichi Wakata, who had arrived at the station 11 days earlier on the STS-119 mission. Unlike previous expeditionary tours, Expedition 19 would merge with Expedition 20 to form the full six-person long-stay crew, but as with recent missions, astronauts and cosmonauts would come and go for shorter periods, visiting and sharing station activities with their long-stay hosts.

Unfortunately, shortly after arriving at the ISS morale suffered for a while from a breakdown in shared acceptance of common needs on the station, as NASA issued a directive to the crews limiting the use of US toilet and exercise machines by cosmonauts. It all resulted from NASA's irritation over the procession of space tourists paying Russia to use common facilities aboard what was an international government laboratory. The result was an agreement imposed from the ground not to share these pieces of equipment or food but this had little relevance to the occupants of the ISS, who had a different view of how micro-communities in space should behave!

Expedition 19 merged with Expedition 20 when Soyuz TMA-15 arrived at the ISS on 29 May. Wakata returned aboard *Endeavour* in July and was replaced by Tim Kopra. Padalka and Barratt returned to Earth aboard TM-14 on 11 October 2009 at 4.32am GMT, formally ending their Expedition 19/20 hybrid mission.

Soyuz TMA-15/ISS 19S Expedition 20

27 May 2009, 10.34am GMT
Cdr: Roman Romanenko (Russia)
Flight Engineer: Frank De Winne (ESA)
Flight Engineer: Robert Thirsk (Canada)

With the arrival of TMA-15 at the ISS on 29 May, the time when the ISS began a six-person operation can be said to have begun when it docked at 12.34pm GMT and its three-man crew joined Wakata, Barratt, and Padalka. But in little more than two years the ISS complement would be reduced due to the Shuttle being retired and a technical failure to a Soyuz launch vehicle. In the interim too there would be various times when the complement slipped below six. Expedition 19 and 20 merged with the arrival of the full resident crew and would have existing flight engineer Wakata returned in two months and replaced with Kopra, who in turn would be replaced by Stott, flown aboard *Discovery* at the end of August.

Padalka and Barratt conducted two spacewalks; the first, lasting 4hr 54min on 9 June, prepared Zvezda for Russia's Poisk airlock/docking module, installed a new antenna, and took images of the Strela-2 crane. The second, on 10 June, lasted 12min and consisted of an internal depressurisation of the Zvezda transfer compartment, technically an

EVA, to install a docking cone for Poisk, which arrived to dock with the aft end of the ISS two days later.

Expedition 20 crewmembers Padalka and Barratt ended their stay on 11 October 2009, when they returned aboard TMA-14 with short-stay Canadian space tourist Guy Laliberté. TMA-15 landed back on Earth at 7.17am GMT on 1 December 2009, with Romanenko, De Winne, and Thirsk on board at the end of their six months aboard the ISS.

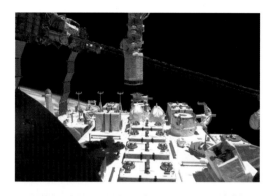

RIGHT The external platform attached to Kibo extended the science tasks Japan could conduct at the station and greatly expanded the unique nature of this component of the ISS. *(JAXA)*

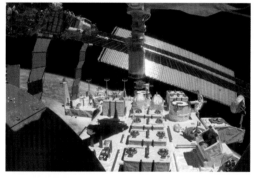

RIGHT The manipulator arm moves down to remove launch restraints and covers from science experiments attached to the unpressurised section of Kibo. *(NASA)*

STS-127/ISS 2JA/Kibo EF and ELM-ES/LCC-VLD1

Endeavour
15 July 2009, 10.03pm GMT
Cdr: Mark Polansky (2)
Pilot: Doug Hurley
MS1: Christopher Cassidy
MS2: Julie Payette (Canada) (1)
MS3: Tom Marshburn
MS4: Dave Wolff (3)
MS5 (up): Tim Kopra
MS5 (down): Koichi Wakata (Japan) (2)

The third launch of dedicated payloads on behalf of the Japanese Aerospace Exploration Agency (JAXA) lifted two crucial elements to the ISS – the Kibo Exposed Facility (EF) and the Kibo Experiment Logistics Module-Exposed Section (ELM-ES). These were the last major pieces of hardware carried to the station and completed the installation of modules that would be used by the partner countries to conduct scientific research. In addition, Wolff would be dropped off with Expedition 20 and Wakata would be carried back to Earth, having been on board the ISS since March.

When *Endeavour* docked to the station on 17 July, the crew joined the six people already in the ISS to bring to 13 the total number of people in space at the same time – a record that still stands and one unlikely to be beaten in the foreseeable future. It also marked the first time two Canadian astronauts were in space at the same time. In addition to the Kibo cargo, *Endeavour* also carried an Integrated Cargo Carrier (see also STS-96) stocked with logistical stores and six new batteries for the P6 truss segment. Attached to the Orbiter docking module was a Light Detection And Ranging (LiDAR) ranging system of the type that would be used by the unmanned SpaceX Dragon commercial cargo carrier, the first test flight of which took place in December 2010.

LEFT Tests with a Dragon Eye sensor that commercial company SpaceX developed for rendezvous and docking of its Dragon logistics capsule is conducted aboard the Shuttle, the sensor here outlined in red. *(NASA)*

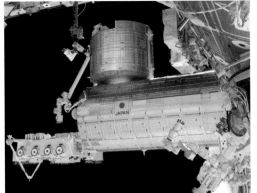

FAR LEFT Astronauts from STS-127 prepare Japan's Exposed Facility on the Kibo laboratory module. *(NASA)*

LEFT The full suite of Japan's Kibo facility as completed with the delivery of the Exposed Facility in July 2009. *(NASA)*

Wolff, Kopra, Marshburn, and Cassidy were involved at various times in five spacewalks between 18 and 27 July, which totalled 30hr 30min. They installed the Kibo elements, moved ORUs from the Integrated Cargo Carrier, replaced six batteries in the P6 truss, and carried out a wide range of maintenance duties. A total 24,638lb (11,176kg) of cargo was transferred to the ISS and 10,479lb (4,753kg) was returned in *Endeavour*, most of that equipment carried up to help install the Kibo elements and carry ORUs for the P6 segment. The Shuttle remained docked to the ISS for 10 days 23hr 41min and returned on 31 July at an elapsed time of 15 days 16hr 44min, having exchanged Wakata for Kopra with the Expedition 20 crew.

Anatomy

Kibo Exposed Facility (EF)

Length	17ft (5.2m)
Width	16.4ft (5m)
Height	12.5ft (3.8m)
Weight	9,000lb (4,082kg)
Power provided	11kW
Power for payloads	10kW at 120VDC
Data management	16-bit 100Mbps

Designed for a life in excess of ten years, the EF is a multipurpose platform physically supporting a range of scientific experiments that can be changed or moved around by the robotic system attached to the Pressurised Laboratory Module (PLM) launched in May 2008. It supports 12 payload attach mechanisms, each of which can support a single experiment as well as the ELM-ES (see below). The EF has power, cooling, data handling, and communication subsystems enabling each experiment to be fully connected to the primary resources and utilities of the ISS.

The Exposed Facility is connected to the laboratory module by a special berthing mechanism incorporating four capture latches each with a motor-driven bolt for structurally attaching it to the exterior of the PLM. The passive part of the mechanism is located on the EF, the active mating component on the PLM, but each has covers removed in orbit during an EVA prior to installation. As with many elements comprising the ISS, Orbital Replacement Units (ORUs) are provided to back up tracking systems, thermal control, and electrical and communication equipment. ORUs fall into separate categories for manual or robotic change-out.

Electrical power is routed via a special exchange unit and can be delivered to any payload attached to the EF, including the HTV

BELOW The unpressurised platform attached to the exterior of Kibo carried two separate remote manipulator systems designed to move emplace, move around or retrieve experiment packages placed outside. *(NASA)*

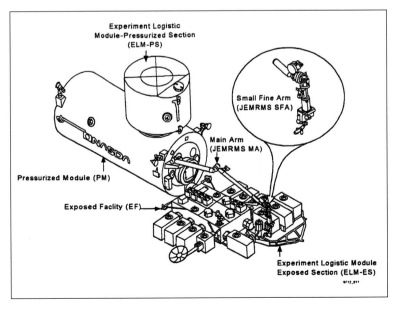

Experiment Logistic Module-Pressurized Section (ELM-PS)

Small Fine Arm (JEMRMS SFA)

Main Arm (JEMRMS MA)

Pressurized Module (PM)

Exposed Facility (EF)

Experiment Logistic Module Exposed Section (ELM-ES)

Exposed Pallet. The equipment exchange unit is designed to allow a wide variety of payloads to be connected to the main operations of the Kibo complex and to have growth potential for future hardware. The EF also carries two sets of video equipment including lights, camera, pan/tilt unit, and associated ORUs, all stored away for launch and set up during an EVA.

Kibo Experiment Logistics Module-Exposed Section (ELM-ES)

Length	13.5ft (4.1m)
Width	16.1ft (4.9m)
Height (including payloads)	7.2ft (2.2m)
Weight dry	2,650lb (1,202kg)
Payload accommodation	3
Power supply	1kW at 120VDC

BELOW The Kibo Small Fine Arm manipulator was designed to make micro-adjustments in movement and alignment. (JAXA)

BELOW RIGHT The Kibo control system is located in the Pressurised Experiment Module and is used to coordinate activities in all three sections of the Japanese segment. (JAXA)

Essentially a platform to which experiment loads can be attached, launched, and returned in the Shuttle, the ELM-ES was a carrier for JAXA payloads. Three were carried aloft on STS-127 and each included a cluster of payloads. A payload interface unit allowed the ELM-ES to be temporarily attached to the Exposed Facility to ease the transfer of equipment off the ELM-ES to the EF proper. Payloads attached to an attachment mechanism could be moved using the Kibo robotic arm connected to the Pressurised Laboratory Module and for this first upload it consisted of several packages.

One experiment monitored X-ray spectra using two special detectors. Another was instrumented to monitor and measure the space environment and its impact on a wide range of materials. A third package was scheduled for the first HTV flight and would construct a global map of stratospheric trace gases to determine the population density of ozone-depleting molecules.

Unlike other partners, Japan installed its own direct-link communication system carrying data, voice, images, and telemetry between the various JAXA modules and the ground station at Tsukuba in Japan. It comprised an antenna, amplifiers, and an inertial reference unit together with Earth and sun sensors.

Integrated Cargo Carrier-Vertical Light Deploy 1

Length	8ft 9in (2.7m)
Width	13ft 9in (4.2m)
Depth	10in (25cm)
Weight empty	2,645lb (1,200kg)
Weight, upload (STS-127)	8,330lb (3,778kg)
Weight, download (STS-127)	6,017lb (2,729kg)

One of only two Integrated Cargo Carriers launched in support of ISS operations, VLD1 was capable of being lifted from the Shuttle cargo bay and positioned so that equipment and ORUs attached to it could be removed and located appropriately, the carrier being returned empty to the Shuttle. VLD2 was flown aboard STS-132.

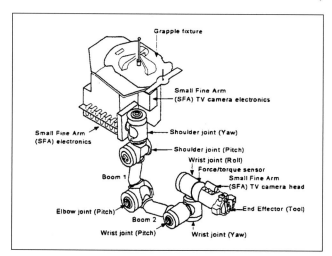

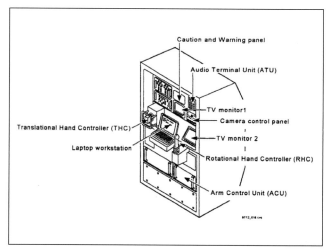

LEFT Containing six new batteries for the P6 truss, spare line-replaceable units, a new pump module and several spares, the ICC-VLD carrier is moved across from the Shuttle to its location on the ISS. (NASA)

Manufactured by Astrium, the VLD provides electrical power and heaters for the payloads and on this mission it carried six batteries for the P6 solar arrays. Each battery ORU contained 38 nickel hydrogen battery cells, or pressure vessels, with two ORUs linked to provide a 76-cell battery pack. Each battery is designed to provide at least 25amps to a maximum 75amps for short-term power demands at a voltage range of 76–123VDC. Together the six batteries weigh 2,204lb (1,000kg) and have a minimum design life of 6.5 years.

Other equipment carried by VLD1 included three Flight Releasable Attachment Mechanism (FRAM) ORUs including the Space-to-Ground Antenna (SAGANT), the Pump Module Assembly (PMA) and the Linear Drive Unit (LDU). SAGANT increases the communications potential of the ISS by providing a back-up Ku-band link with the Tracking & Data Relay satellite System (TDRS) conveying payload data, video, and crew communications including emails and telephone calls. The antenna dish is 6ft (1.8m) in diameter and 6ft high. It is mounted on gimbals and weighs 194lb (88kg).

The PMA is an integral part of the station's Active Thermal Control System, which provides cooling to internal and external avionics, payloads, and the crew using ammonia pumped through loops by the PMA. It weighs 780lb (354kg) and is 5ft 9in long by 4ft 2in wide and 3ft high (1.75m x 1.27m x 0.9m). Built by Northrop Grumman, the LDU provides motive power for the Mobile Transporter mounted on truss rails and features two independent drive and engagement systems. It is 4ft wide, 3ft high and almost 2ft deep (1.2m x

0.9m x 0.6m), with a weight of 255lb (116kg). With the exception of the batteries, VLD1 ORUs were stored on ESP-3 on the P3 truss.

STS-128/ISS 17A

Discovery
28 August 2009, 3.59am GMT 29 August
Cdr: Rick Sturckow (3)
Pilot: Kevin Ford
MS1: Patrick Forrester (2)
MS2: Jose Hernandez
MS3: Danny Olivas (1)
MS4: Christer Fuglesang (ESA) (1)
MS5 (up): Nicole Stott
MS5 (down): Tim Kopra

The mission of STS-128 was to install science and system racks, carried up in the Leonardo MPLM along with the Combined Operational Load-Bearing External Resistance

BELOW In August 2009 *Discovery* delivered a wide range of supplies inside the Leonardo Multi-Purpose Logistics module, one of a series of logistics modules built in Europe. (NASA)

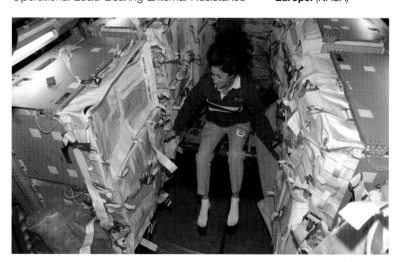

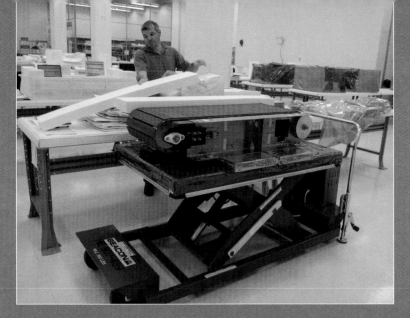

ABOVE Known as COLBERT, a sophisticated fitness treadmill was carried to the station in the Leonardo MPLM, here seen being packed for flight. *(NASA)*

ABOVE The COLBERT fitness machine was named after comedian Stephen Colbert who featured on this dedicated decal. *(NASA)*

COLBERT AND THE FITNESS REGIME

BELOW Physical fitness is a vital necessity for resisting the potentially harmful effects of weightlessness and the TVIS seen here was another basic exercise device worked strenuously by all crewmembers. *(NASA)*

The Combined Operational Load-Bearing External Resistance Treadmill was designed to examine the degree of physical adaptability necessary for future long-duration trips into deep space. It was the second treadmill on the station, the Treadmill with Vibration Isolation Stabilisation (TVIS) being the first, and one of six devices to compensate the body for the effects of weightlessness. Unlike the TVIS which had been developed at NASA, the treadmill base for COLBERT was procured from Woodway USA in Wisconsin and fitted with supporting equipment such as power, cooling, and other structures for use in zero-g. The device was carried to the ISS in an International Standard Payload Rack (ISPR) and stowed in Node 2 (Harmony) until eventually being relocated to Tranquility (Node 3), launched in February 2010.

A significant difference between the TVIS and COLBERT is that whereas the former uses an active system that senses running forces and 'throws' a counterbalancing mass to compensate, and thereby reduce vibrations, the latter has inherent mass to maintain stability. With a standard launch weight of 1,600lb (726kg), when fitted for duty COLBERT weighs 2,200lb (998kg) but the entire rack has a passive isolation system using springs to dampen vibration. Although more massive, COLBERT is simpler and potentially longer-lasting. With crewmembers needing to work

out each day to counteract flagging muscles and bone demineralisation, a reliable system is essential to healthy conditioning.

Each astronaut or cosmonaut must work out for a minimum of two and a half hours each day, of which almost half will be on one of the two treadmills. In that time they are expected to work off 250–500 calories, and COLBERT has instrumented load cells and accelerometers to measure the amount of foot-lb force used during these sessions. Using real-time data, physicians evaluate the degree of effectiveness in counteracting the effects of zero-g and adjust the regimes accordingly. In the absence of gravity, bungee cords restrain the user in a vertical position and apply loads against which the muscles work, adjustable to a prescribed level for each individual.

ESA is developing a system that will more effectively simulate 1-g and provide a higher fidelity in body toning – as important for experiencing g-forces on return to Earth as gathering data for long-duration flight. In a bizarre way, the crewmembers give something back to the station. In taking in food and combusting it to produce energy, they perspire and add moisture to the station's environmental system, which uses this fluid to recirculate water into the operating system! COLBERT was named after the TV comedian when astronaut Suni Williams 'ran' the Boston Marathon in space.

Treadmill (COLBERT), named after the TV comedian Stephen Colbert. The MELFI-1 freezer carried aboard STS-121 and installed in Destiny had subsequently been moved into the Kibo laboratory so the MELFI-2 freezer lifted up by *Discovery* on this flight was moved directly from the MPLM to Destiny. Leonardo also carried up the Node 3 Air Revitalisation System Rack temporarily placed in the Kibo pressure module until Node 3 was launched. It would remove carbon dioxide from the ISS, remove potentially harmful trace elements, and monitor the cabin atmosphere for various chemical elements.

The Orbiter also carried a Lightweight Multipurpose Experiment Support Structure Carrier (LMC) weighing 3,926lb (1,781kg), first flown on STS-108 and on three subsequent missions. For this flight Leonardo had a weight of 27,510lb (12,478kg) at launch, returning with 16,268lb (7,387kg). Leonardo carried two research racks, four station system racks, seven Resupply Stowage Platforms, two Resupply Stowage Racks, one Zero Stowage Rack, and one Integrated Stowage Platform. Modifications to the aft cone assembly allowed 12 extra cargo bags – each the size of a carry-on suitcase – with additional lithium hydroxide CO_2 removal canisters and Remote Power Control Modules for the electrical power system.

In addition to the two crew quarters installed in Node 2, the new two-person quarters brought up by *Discovery* were installed in the Kibo module, each providing private space and soundproofing, better airflow, redundant electrical systems, and caution and warning equipment. The size of a standard rack, the crew quarters have adjustable lighting, selectable ventilation, and fixtures to allow each crewmember to personalise their space. The two science racks consisted of a Fluids Integrated Rack (FIR) and a Materials Science Research Rack (MSRR-1). The FIR was for

FAR LEFT The first ESA astronaut to command the space station Frank De Winne led Expedition 21 and helps install new two-person crew quarters in the Kibo Pressurised Laboratory Module. *(ESA)*

FAR LEFT Working with new materials and experimenting with the effects of weightlessness on others is a vital part of what scientists call 'microgravity research', a task pursued by the MSRR-1 carried to the ISS aboard *Discovery* in August 2009. *(NASA)*

LEFT The low-gradient furnace for the European-built MSRR-1 materials science laboratory, to be located in Destiny. *(NASA)*

research into colloids, gels, bubbles, wetting, and capillary action under microgravity conditions to better understand how fluids behave. This research would show how fluid-carrying vessels such as fuel tanks and life support systems can be designed to operate better in these unique conditions. MSRR-1 is essentially a facility compatible with a wide range of experimental investigations into the way materials behave. Metals, alloys, polymers,

semiconductors, ceramics, crystals, and glasses can all be studied with this equipment, enabling a wide and diverse range of studies over long or short periods. It houses the ESA Materials Science Laboratory and was located in the Destiny module.

Discovery docked with the PMA-2 port on 30 August and Leonardo was moved across to the nadir port on Harmony the following day so that offloading could begin. Three spacewalks were undertaken lasting 19hr 15min, one of which included replacement of the ammonia tank used in the ISS Thermal Control System. In all, the Shuttle transferred 18,548lb (8,413kg) to the ISS with 5,223lb (2,369kg) returned in *Discovery*. At the end of the mission, the ISS had a mass of 710,966lb (322,489kg). With Leonardo back inside *Discovery*'s cargo bay on 7 September, the Shuttle undocked after 8 days 18hr 32min and landed at Edwards Air Force Base, California, on 11 September, at an elapsed time of 13 days, 20 hours and 54 minutes.

HTV-1/ISS-HTV1 Kounotori 1

10 September 2009, 5.02pm GMT

Launched by Japan's powerful H-IIB launcher from the picturesque Tanegashima launch site, HTV-1 carried the Superconducting Submillimetre-Wave Limb Emission Sounder (SMILES) and an experiment payload in the Unpressurised Logistics Carrier for placement

on the Kibo external pallet. In the Pressurised Logistics Carrier was 7,938lb (3,600kg) of cargo for the ISS of which one-third was food, about a fifth was laboratory experiments. The rest constituted robotic arm hardware for Kibo, crew supplies and personal items, packaging materials, and clothing. The flight marked the first launch of the H-IIB rocket specifically developed for the HTV and the first ascent from the second pad at the Yoshinobu complex.

Launched to an interim orbit of 162nm x 108nm, HTV-1 conducted its own rendezvous manoeuvres and was within 3nm of the ISS at 1.59pm GMT on 17 September, ready for the final approach that began at 3.31pm GMT that day. Closing to a distance of 33ft (10m) it was grasped by Canadarm2 under the control of Nicole Stott at 7.47pm GMT, with final lock-on being achieved four minutes later. Robert Thirsk then used Canadarm2 to move it into position over the Harmony nadir port at 10.08pm GMT, with capture four minutes later.

HTV-1 remained docked for six weeks, undocking at 3.02pm GMT on 30 October when Canadarm2 moved it away from the Harmony module to a position about 38ft (11.6m) below the ISS, releasing it at 5.32pm GMT. From there it was commanded to destructive re-entry through a series of engine burns across a region of the atmosphere over the Pacific Ocean between New Zealand and South America. On board were 199 items of waste transferred from the ISS of which 1,605lb (728kg) was discarded rubbish and 1,975lb (896kg) comprised empty racks, a total of 3,580lb (1,624kg).

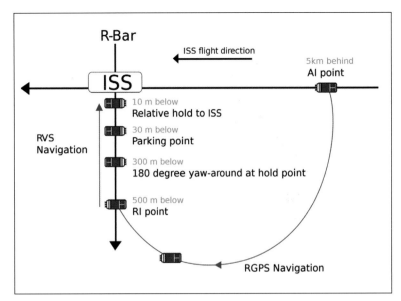

Anatomy

H-II Transfer Vehicle

Length	30.2ft (9.2m)
Length two PLC	24.3ft (7.4m)
Diameter	14.5ft (4.4m)
Weight empty	23,152lb (10,500kg)
Launch payload max	13,230lb (6,000kg)
Pressurised payload	11,466lb (5,200kg)
Unpressurised payload	3,307lb (1,500kg)
Maximum weight at launch	36,382lb (16,500kg)

Japan has had a long history of launching its own satellites and developing its own rockets

and launch vehicles. Its commitment to the ISS was one of the largest outside the US and an additional investment was made in designing and constructing the unmanned H-II Transfer Vehicle, so named because it is launched by Japan's H-IIB rocket specially developed from the H-II for launching the HTV logistics module. In function it is similar to ESA's Automated Transfer Vehicle (ATV) but now that the Shuttle has been retired it is the only logistics module

ABOVE Japan's HTV used the R-bar approach method whereby the chase vehicle (HTV) maintains a constant pitching manoeuvre to close on the target (the ISS) from below in its circular orbit of the Earth. *(NASA)*

LEFT Grasped by the station's manipulator, Japan's HTV-1 is docked to the Harmony nadir (Earth facing) port. *(NASA)*

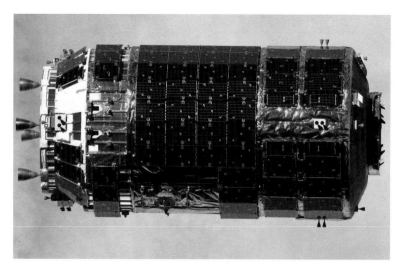

capable of carrying the International Standard Payload Rack (ISPR). The HTV is named Kounotori, which in Japanese means 'white stork', to denote the arrival of new things and the birth of new life for the Kibo module. Development of the HTV began at Japan's National Space Development Agency in the early 1990s and was developed under the new agency combining all Japanese aerospace research and development, JAXA.

The HTV comprises four cylindrical sections attached together during manufacture: the

Pressurised Logistics Carrier (PLC), the Unpressurised Logistics Carrier (UPLC), the Avionics Module (AM), and the Propulsion Module (PM). The PLC carries supplies that will be used aboard the ISS and allows crewmembers to work in a shirtsleeve environment. It docks directly to the Harmony module by a Common Berthing Mechanism (CBM). The UPLC carries unpressurised equipment and is immediately aft of the PLC. It can be accessed by a robotic arm and carries pallet payloads for installation on the Kibo Exposed Facility aft of the Kibo module. Further aft of the UPLC, the Avionics Module contains all the navigational and electrical equipment needed for autonomous operation of the HTV and for communications and data relay. The Propulsion Section is furthest aft of the other sections and contains the propellant tanks, pressurisation systems and rocket motors for orbital rendezvous and attitude control.

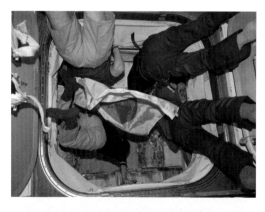

RIGHT With the Japanese flag fluttering in a weightless breeze from air conditioning, the HTV-1 module is opened for access. (NASA)

A major departure for the HTV is that it does not have independent final approach and automated docking and once it has positioned itself close to the Harmony module, Canadarm2 on the Mobile Transporter is used to grasp the HTV and berth it at a docking port, hard lock being achieved with 16 bolts. The logistical supplies it can carry include eight full International Standard Payload Racks with two on each of the four faces of the Pressurised Logistics Carrier. The baseline version of the HTV carrying mixed cargo can be changed during pre-flight preparation to replace the Unpressurised Logistics Carrier with another Pressurised Logistics Carrier to allow twice the crew access volume, in which event the overall length is reduced somewhat, as the UPLC is longer than the PLC.

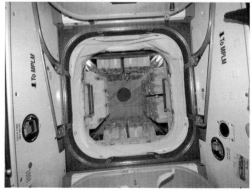

RIGHT Although the signs on the Harmony 'walls' indicate the way to a European-built MPLM, for this docking the received vehicle is Japan's logistics module. (NASA)

The Propulsion Module includes four main engines, each with a thrust of 112lb (51kg), and 28 thrusters, each of 25lb (11kg) thrust using bi-propellant monomethyl hydrazine (MMH) and mixed oxides of nitrogen MON_3 as propellants. The main thrusters are manufactured by Aerojet and are essentially the same as those previously used as vernier motors on the Shuttle. The smaller thrusters are R-4D heritage, dating back to the days when they were used on the Apollo Service Module. The Propulsion Module carries 5,290lb (2,400kg) of propellant in four tanks.

Soyuz TMA-16/ISS 20S Expedition 21/22

30 September 2009, 7.14am GMT

Cdr: Maksim V. Suraev (Russia)
Flight Engineer: Jeffrey N. Williams (US)
Tourist: Guy Laliberté (Canada)

The next set of overlapping long-stay crews, Expedition 21 was unique from the outset as the first in which three Soyuz vehicles would be docked to the ISS at the same time: TMA-14, -15 and -16. It carried Canada's first space tourist, Guy Laliberté, who returned home in TMA-14 after a flight lasting 10 days 21hr 17min, along with the resident Expedition 20 crew of Padalka and Barratt. Because of the retirement of the Shuttle and reliance on Soyuz spacecraft for carrying expeditionary crews back and forth, Laliberté was the last such space tourist. Further opportunities will only open up once again when US commercial or NASA spacecraft begin flights to the ISS.

Expedition 21/22 set up new research facilities in materials and fluid sciences and operated the COLBERT treadmill lifted up the previous December. In all they would receive the Japanese HTV-1 module, help integrate the Poisk docking/airlock module, and host two Shuttle crews and a Progress cargo-tanker.

Padalka, Barratt, and Laliberté returned to Earth at 4.32am on 11 October 2009. After De Winne, Romanenko, and Thirsk returned home in Soyuz TMA-15, landing at 7.17am GMT on 1 December, Williams assumed command of the station under Expedition 22. Suraev and Williams landed back on Earth in TMA-16 on 18 March 2010 after a space-flight time of 169 days 4hr 10min.

Progress M-MIM2/ISS-35P Poisk

10 November 2009, 2.22pm GMT

In further exploitation of the instrument and propulsion section of the Soyuz spacecraft, like Pirs launched in 2001, Poisk was lifted to the ISS by the aft component of the Progress M cargo-tanker. With a weight of 15,660lb (7,100kg) including the Poisk module, M-MIM2 (originally M-SO2) was launched to an interim orbit by a Soyuz-U rocket to join with the ISS and the Expedition 21 crew. Following a standard final approach using the Kurs system, the docking interface on Poisk latched on to the Zvezda zenith port at 3.41pm GMT two days later, and within several minutes a hard-dock had been achieved – just as the ISS was flying over Kazakhstan.

ABOVE Like Pirs before it, Russia's Poisk module incorporated both docking and airlock facilities and was lifted to the Zvezda zenith (space-facing) port. *(NASA)*

FAR LEFT The interior of Poisk is capable of supporting two suited space walking astronauts from where they will egress and ingress for work outside the station. *(NASA)*

LEFT Having done its job, the propulsion section adapted from the Soyuz/Progress family of spacecraft is jettisoned for de-orbit. *(NASA)*

Access to Poisk was gained by Romanenko and Suraev when the hatch was opened at 12.17pm GMT on 13 November. A troubling but erroneous alarm indicating rapid decompression, thought to have originated in the Poisk module, sent the crew scurrying on two occasions. The propulsion and instrument section separated from Poisk at 16 minutes after midnight GMT on 8 December and a 38-second de-orbit burn at 4.48am GMT began its fiery demise in the atmosphere.

Anatomy

Poisk

Length	13.3ft (4m)
Diameter	8.4ft (2.6m)
Weight	8,092lb (3,670kg)
Pressurised volume	523ft^3 (14.8m^3)
Habitable volume	378ft^3 (10.7m^3)
Deliverable cargo	2,205lb (1,000kg)

At first known as Docking Module 2 and then Mini-Research Module 2, the Poisk airlock and docking module is similar to the Pirs system launched earlier and is designed to expand the opportunities for combined docking and EVA facilities from the Zvezda research module. It provides space for scientific experiments and facilities for power supply and data transmission between Zvezda and the modules docked to it. Poisk also supports a cargo boom attached to one side of the module, adjacent to one of the two EVA hatches, with locations for scientific experiments on the exterior for exposure to the vacuum of space.

Designed and built by S P Korolev RSC Energia Corporation, Poisk has two opposing EVA hatches, each of 3ft (91cm) diameter with a single docking interface. Located on the zenith port of Zvezda, Poisk completes the combined docking module/airlock module update and adopts much of the technology originally planned for the Mir-2 station.

STS-129/ISS ULF3

Atlantis
16 November 2009, 7.28pm GMT
Cdr: Charles O. Hobaugh (2)
Pilot: Barry E. Wilmore
MS1: Leland Melvin (1)
MS2: Randy Bresnik
MS3: Mike Foreman (1)
MS4: Robert Satcher Jr
MS5 (down): Nicole Stott

With 38,893lb (17,642kg) of equipment in the cargo bay and 29,458lb (13,362kg) going up to the station, STS-129 was a 'stock and store' flight packed with ORUs, a cryogenic storage freezer known as GLACIER, biological research equipment, replacement parts, and two large ExPRESS Logistics Carriers for attachment to the truss assembly. The first of these, ELC-1, was lifted from the cargo bay and handed over to Canadarm2 which attached it to the P3 truss. ELC-2 was lifted to its position on the S3 truss segment on 21 November, just before the second of three spacewalks.

Three EVAs totalled 18hr 27min during which Satcher, Foreman, and Bresnik had various excursions installing a new antenna on the truss, deploying fixtures and fittings brought up by the Shuttle, installing a grapple bracket to Columbus, relocating some ORUs, installing a high-pressure gas tank on Quest, and conducting general maintenance duties.

Bresnik was the on-orbit custodian of a scarf which had been worn by the famous American aviator Amelia Earhart, an item formerly on display at the Ninety-Nines Museum of Women Pilots in Oklahoma City, to which it was returned after the STS-129 mission. *Atlantis* remained docked to the ISS for 16 minutes shy of six days and separated on 25 November, to return home two days later at an elapsed duration of 10 days 19hr 16min.

BELOW NASA's Jeffrey Williams checks out the newly arrived Poisk module, with its two opposing hatches for space walks. *(NASA)*

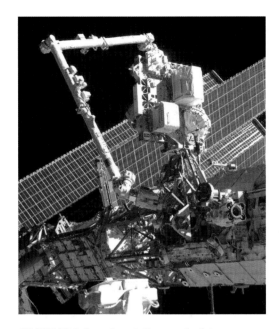

ABOVE ELC-2 on the station manipulator arm, being moved from *Atlantis* to the S3 truss segment. *(NASA)*

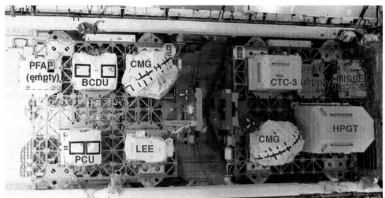

ABOVE ELC-1 and ELC-2 with equipment including: BCDU (Battery Charger Discharge Unit), CTC (Cargo Transportation Container), CMG (Control Moment Gyroscope), PFAP (Passive Flight Releasable Attachment Mechanism), LEE (Latching End Effector), HPGT (High Pressure Gas Tank), PCU (Power Control Unit) and MISSE (Materials International Space Station Experiment). *(NASA)*

Anatomy

ExPRESS Logistics Carrier (ELC)

Length	16ft (4.9m)
Width	14ft (4.3m)
Weight ELC-1	13,850lb (6,282kg)
Weight ELC-2	13,400lb (6,078kg)

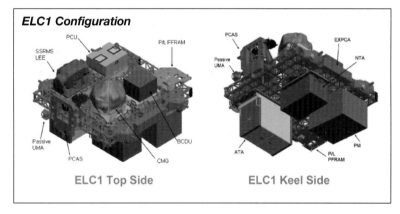

ELC1 Configuration

ELC1 Top Side   ELC1 Keel Side

ABOVE A logistics 'stock and store' flight by *Atlantis* in November 2009 packed more than 19 tons into the Shuttle's cargo bay, of which almost 15 tons were in supplies left at the ISS. *(NASA)*

BELOW The two ExPRESS Logistics Carriers uplifted to the ISS by *Atlantis* on STS-129 carried a wide range of experiments and sundry equipment. *(NASA)*

Designed and built by NASA's Goddard Space Flight Center, and based on their experience with cargo carriers for Hubble Space Telescope upgrades, four ELC units were on the Shuttle manifest for delivery before the last flight. The name ExPRESS is an acronym for 'Expedite the Processing of Experiments to the Space Station'. STS-129 was the first flight for the ExPRESS carriers, the remaining ones being ELC-4 launched by *Discovery* (STS-133) in February 2011 followed by ELC-3 aboard *Endeavour* (STS-134) in May 2011.

ELC-1 was stored on the UCCAS mounting at the P3 truss segment with ELC-2 on the PAS unit at the S3 segment. Each is an unpressurised platform capable of supporting 9,800lb (4,445kg) of equipment for general engineering requirements or for scientific experiments. The ELC provides not only a platform for attachment but also electrical

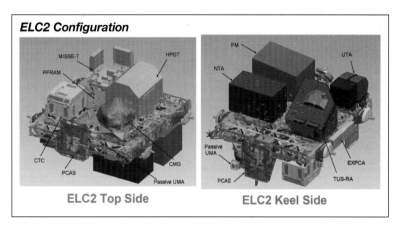

ELC2 Configuration

ELC2 Top Side   ELC2 Keel Side

power, communications, data flow, and control systems. Each can carry 12 Flight Releasable Attachment Mechanism (FRAM) cargo modules, each of which has two payload attachment sites with full avionics provided.

The ELC is built to have direct interface with the ISS integrated truss common attach system, two positions being on the P3 truss at zenith and nadir positions, and four positions on the S3 segment, two facing zenith and two facing nadir. The ELC offers scientists the opportunity to fly experiments at less cost than the procurement of a dedicated satellite, with power and data transmission to the ISS and thence to the ground.

The electrical system provides power distribution through two 3kW 120vdc feeds and the avionics suite provides a ColdFire based flight computer, a microprocessor derived from the Motorola 68000 architecture. ELC-1 and -2 carried 14 large ORUs, the largest number of orbital replacement spares lifted thus far, including an ammonia tank assembly for the cooling system, a cargo transportation container, two Control Moment Gyroscopes (CMGs), a high-pressure gas tank, a latching end effector for Canadarm2, experiment containers, nitrogen tank assemblies, a Pump Module Assembly (PMA), and a trailing umbilical reel assembly for the Mobile Transporter.

As further applications in the MISSE

(Materials ISS Experiment) programme, MISSE-7 samples were carried in two experiment trays within two Passive Experiment Containers (PECs) which were opened in orbit to expose samples to the extremes of the space environment. MISSE-7 included electronic samples provided by the Naval Research Laboratory, the Air Force Laboratory and NASA.

Soyuz TMA-17/ISS 21S/ Expedition 22/23

20 December 2009, 9.52pm GMT
Cdr: Oleg Kotov (Russia)
Flight Engineer: Timothy Creamer (US)
Flight Engineer: Soichi Noguchi (Japan)

The first December launch of a Soyuz in 19 years – a month avoided if possible due to poor weather conditions in abort and recovery areas – saw the crew dock to the Zarya nadir port. On 12 May 2010, however, TMA-17 was moved to the aft Zvezda port, making way for the Rassvet module. The crew joined Suraev and Williams aboard the ISS but when their Expedition 22 mission ended on 18 March they became the resident crew for Expedition 23. TMA-17 returned to Earth with Kotov, Creamer, and Noguchi on 2 June 2010, landing in Kazakhstan at 3.25am GMT.

STS-130/ISS 20A/ Tranquility Node 3 and Cupola

Endeavour
8 February 2010, 9.14am GMT
Cdr: George D. Zamka (1)
Pilot: Terry W. Virts Jr
MS1: Kathryn P. Hire (1)
MS2: Stephen K. Robinson (3)
MS3: Nicholas J. M. Patrick (2)
MS4: Robert L. Behnken (1)

The primary purpose of this mission was to deliver Node 3 (Tranquility) and the Cupola. There had been a long campaign to provide the ISS with a viewport through which crewmembers could get an unprecedented view of the Earth below. For that reason it

BELOW MISSE-7 being prepared for flight aboard Atlantis to the station on STS-129. *(NASA)*

would be attached to the Earth-facing side of the station but it would also serve to provide near-panoramic views of the truss and robotic equipment on the exterior. After docking at 5.26am on 10 February, the crew joined Suraev, Williams, Kotov, Creamer, and Noguchi aboard the ISS and began preparations for the first spacewalk, involving Patrick and Behnken.

In a 6hr 32min EVA they presided over the transfer of Tranquility, with the Cupola attached to one end, to the port side of Unity (Node 1) where it was attached using the station's robotic arm and accessed by the crew the following day to check it out. The second EVA – on 14 February – lasted 5hr 54min, during which Behnken and Patrick conducted a wide range of installation tasks, including outfitting the nadir port on Tranquility to receive the Cupola and installing additional equipment.

The next day the Cupola was lifted off the end of Tranquility and relocated to its new Earth-facing position using Canadarm2. On 16 February PMA-3 was moved from the zenith position on Node 2 (Harmony) to the end docking unit on Node 3 to which future Shuttles could dock. The third and final spacewalk took place on 17 February. In a 5hr 48min spacewalk Behnken and Patrick connected cables to PMA-3, removed covers and launch locks from the Cupola, and allowed the window shades to be opened for the first time in space.

LEFT Parallel to Japan's Kibo module, Node 3 gets attention as astronauts Robert Behnken and British-born Nick Patrick remove flight covers from the Cupola window shades. *(NASA)*

LEFT PMA-3 in the process of being moved from the zenith position on Node 2 (Harmony) to the end of Node 3 (Tranquility), to which the Cupola had been attached for launch. *(NASA)*

LEFT With Tranquility attached to the port side of Node 1 (Unity), and the Cupola now permanently fixed to the nadir (Earth facing) port, PMA-3 has been relocated for future docking operations. *(NASA)*

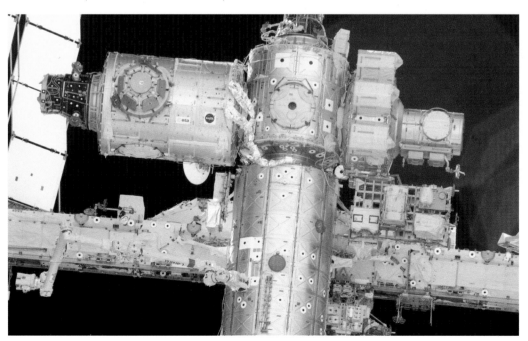

LEFT Attached to Destiny (below), Node 1 (top centre) supports the Quest airlock (right) and Node 3 with the Cupola (left). *(NASA)*

Endeavour undocked at 12.54am GMT on 20 February and landed back at the Kennedy Space Center at 3.22am GMT two days later. The mission had lasted 13 days 18hr 8min.

Anatomy

Tranquility Node 3

Named after the site selected for the first Moon landing, Tranquility was one of two nodes built by ESA for the ISS. Surviving many changes brought to the station since it was

commissioned in 1997, the role and function of Node 3 has metamorphosed from a connecting module to a highly complex facility providing life support functions and a pathway into the Cupola permanently attached to its nadir port. It was only through the addition of this facility that the long-term crew complement could increase from three to six people.

Located on the port side of the Unity (Node 1) module, Tranquility has the same structural design and size as Node 2 (Harmony), with six docking ports but with an installed weight of 39,665lb (17,992kg). The zenith (space-facing) port has been inhibited and is now used to park the Special Purpose Dextrous Manipulator (SPDM). The other two (forward and aft-facing) ports are available for additional ISS modules as necessary.

RIGHT Named Tranquility, after the site for the first Apollo Moon landing, Node 3 looks clean and neat as it arrives on station. *(NASA)*

Unlike the shorter Node 1 to which it is attached, Node 3 can be considered as two modules joined together. The aft section closest to Unity has provision for eight standard equipment racks arranged in four pairs at 90 degrees to each other to form a box structure inside with four flat 'walls'. These racks would eventually carry systems and equipment necessary for the node's functions and operational requirements. The forward section supports the five docking ports – one axial and four radial, also at 90-degree intervals around the circumference of the module. These were designed in to Tranquility when additional elements now cancelled were expected to dock with the module. These included the abandoned US Habitation Module and the Crew Return Vehicle, a mini-shuttle moored to the station for emergency evacuation and return to Earth. That latter function is now provided by Russian Soyuz spacecraft.

RIGHT Jeffrey Williams works the Urine Processor Assembly Distillation Assembly, one of many 'housekeeping' tasks repairing, replacing and generally maintaining vital systems and subsystems. *(NASA)*

Three racks launched to the ISS in Node 3 carried pallets of supplies and equipment and two contained avionics equipment but three rack spaces were empty. The three racks carrying equipment and cargo were returned aboard STS-131 in April 2010 and six racks delivered to the ISS earlier were moved in to Tranquility. These included the second Air Revitalisation System monitoring the quality and chemical composition of the air, an Oxygen Generation System to produce oxygen from water, two Water Recovery System racks for processing urine and water, a Waste and Hygiene Compartment rack for crew waste and personal hygiene, and a second treadmill. Node 3 would also house the Advanced Resistive Exercise Device.

Tranquility did more than bring equipment and provide volume for containing additional systems. It also served as a connection between Nodes 1 and 2 for the distribution of water for Shuttle fuel cells, for drinking and for the transfer of pre-treated urine, and for moving oxygen and nitrogen around the station. The two avionics racks accommodated virtually all the electronic equipment for the command and data handling function and for the conversion and distribution of electrical power from the solar arrays to the entire station. Powerful computers in the avionics racks provide fault isolation and recovery algorithms so that the power distribution system can work efficiently and safely.

Like Node 2 before it, Tranquility was an evolution of the Multi-Purpose Logistics Module and reached greater levels of sophistication than had been envisaged at the outset, containing the most advanced life support system ever flown into space. Built by Italy, Tranquility remained in storage at Thales Alenia Space until it was shipped to the Kennedy Space Center on 17 May 2009, arriving at the end of May, when it was made ready for launch.

Anatomy

Cupola

Height	4.9ft (1.5m)
Diameter	9.5ft (2.9m)
Weight	4,145lb (1,880kg)

A legacy from one of the earliest days of formal space station planning at NASA in 1987, the Cupola was originally envisaged as a place where astronauts could observe and monitor

work conducted by astronauts or robotic arms by physically viewing activity rather than observing it on TV screens. It evolved as an important part of the space station concept, a place where living in space was a fundamental consideration for the sort of environment provided for people remaining in orbit for lengthy periods of time. Astronauts supported the notion that a crew that could observe the Earth through large windows would benefit psychologically and behaviourally.

At various times it was a candidate for cancellation when budget pressures wrought havoc with successive concept plans for the station, but it was finally picked up by ESA as a piece of hardware which Europe would fabricate and provide to NASA as part of its contribution to the ISS. The contract to build the Cupola was signed on 8 February 1999, with fabrication and assembly being the responsibility of Alenia Spazio in Italy. Material contributions also came from companies in Spain, Switzerland, Sweden, Germany, and Belgium.

The design is of a suppressed truncated conical dome with one axial and six radial window locations. Fabricated from aluminium 2219-T851, the dome and skirt assembly carry seven windows in all. One of them, at 31.5in (80cm) diameter placed axially, is the largest window ever flown into space. Six trapezoidal-

ABOVE Long campaigned for – a room with a view. The Cupola provides magnificent views from its perpetually Earth-facing vantage point. *(NASA)*

shaped windows are positioned radially at equal intervals around the circumference. A passive CBM is attached for docking, bolted to the skirt assembly, and a micrometeoroid protection layer carries an aluminium bumper on the cylindrical section. The dome gains structural strength through being a single forged unit with no welds.

The windows are manufactured from fused silica and borosilicate glass and each has four panes including an inner scratch pane, two 0.98in (2.5cm) thick pressure panes, and an external debris pane. Inner and outer panes can be replaced if necessary without compromising the pressure inside the Cupola. Each window has an aluminium 6061-T6 shutter, or eyelid, that can be closed or opened as necessary and each is covered with Kevlar/Nextel for impact protection. The shutters are kept closed when the Cupola is not in use to help protect the windows from impact damage.

Electrical power for the window heaters, robotic work station, portable computer system, and lighting comes directly from the Node 3 120VDC supply, and the communications and data subsystem is tied in to the MIL-1553B data bus via a utility outlet panel. Flight hardware includes two Flight Releasable Grapple Fixture interface units for moving the Cupola. Upper and lower handrails are provided inside the Cupola as well as covered conduits and airways for the distribution of atmospheric gases.

Soyuz TMA-18/ISS 22S/ Expedition 23/24

2 April 2010, 4.04am GMT
Cdr: Aleksandr Skvortsov (Russia)
Flight Engineer: Mikhail Korniyenko (Russia)
Flight Engineer: Tracy Caldwell Dyson (US)

Docking took place at 5.25am GMT on 4 April when TMA-18 latched on to the Poisk module, which had been attached to the Zvezda zenith port the previous November. The new long-duration crew joined Expedition 23 crewmembers Kotov, Creamer, and Noguchi until the latter returned to Earth on 2 June. Remaining aboard to initiate Expedition 24 and receive its long-duration crew on 17 June, TMA-18 was due to separate and depart for Earth

early on 24 September. But problems with the connecting latches prevented that and the three crewmembers had to return to the station while the problem was solved. Undocking finally took place at 10.02pm GMT on 24 September and TMA-18 brought Skvortsov, Korniyenko, and Caldwell Dyson back to land at 5.23am GMT the following morning, 176 days after their launch.

STS-131/ISS 19A

Discovery
5 April 2010, 10.21pm GMT
Cdr: Alan Poindexter (1)
Pilot: James Dutton
MS1: Richard Mastracchio (2)
MS2: Dorothy Metcalf-Lindenburger
MS3: Stephanie Wilson (2)
MS4: Naoko Yamazaki (Japan)
MS5: Clayton Anderson (1)

Primarily a logistics delivery flight, *Discovery* carried the Leonardo MPLM on its last round-trip to the ISS; the next time it flew it would remain docked to the station. It was also the last time a seven-person crew would ride any spacecraft for a very long time and the last Shuttle flight on which at least one of the astronauts was a rookie – the final four Shuttle missions would carry all-veteran crews. This was also the last flight on which three female astronauts would ride into space together and the last time four women were in space at the same time, as the three women from STS-131 joined Tracy Caldwell Dyson on the ISS.

Payload included a Lightweight MPESS Carrier (LMC), stowed at the aft end of the Shuttle's cargo bay, forward of which was the Leonardo module. Although it had flown before, the weight-lifting capacity of Leonardo was improved by removing 178lb (81kg) of unnecessary hardware from the basic structure. It carried 16 racks: four experiment racks, a systems rack, seven Resupply Stowage Platforms, and four Resupply Stowage Racks. In addition, Leonardo had a fully stocked aft cone stowage area housing 12 bags stuffed with supplies, each the size of a carry-on suitcase, a configuration first flown on STS-126.

Leonardo delivered the Window Observational Research Facility (WORF) slanted towards outreach and educational programmes providing images for crop and weather damage analysis. Attached to the interior nadir side of the Destiny laboratory, WORF benefited from that module's purest optical window ever flown and provided views 39 degrees forward, 32 degrees aft, and 79 degrees from port to starboard in a single image. Built around a standard ExPRESS rack, WORF could support various sensors including cameras and camcorders, and carried attachment points for a wide range of sensor applications, aiding teaching and stimulating students.

Also carried to the ISS, the Muscle Atrophy Research and Exercise System (MARES) consisted of an adjustable motorised chair attached to an adjustable pantograph and associated control devices for research into musculoskeletal, biomechanical, and neuromuscular physiology. MARES would measure and exercise seven different limb joints with nine angular movements and two linear movements on the arms and legs. Physicians believe it is considerably more advanced than ground-based dynamometers and that it will produce valuable data for combating ailments suffered by Earthlings. It is integrated into an International Standard Payload Rack known as the Human Research Facility, combining several tests and studies aimed at significantly improving the physical condition of humans in space or on Earth.

Leonardo also carried the third Minus Eighty-Degree Laboratory Freezer (MELFI-3), the first two having been lifted to the station aboard STS-121 and STS-128. Information about MELFI can be found in the payload description for STS-121. Another welcome addition to the ISS was the fourth Crew Quarters (CQ-4), this one to be installed in the Node 2 (Harmony) module. Three EVAs were conducted, all by Mastracchio and Anderson, for a total 20hr 17min during which Canadarm2 was used to stow a new ammonia tank on the ISS, removing the old tank and carrying out a range of maintenance tasks. In all, *Discovery* carried 31,131lb (14,121kg) of cargo and would return to Earth with 21,764lb (9,872kg). After 10 days 5hr 8min docked to the ISS, *Discovery* separated from the station and returned to Earth, touching down at 15 days 2hr 47min.

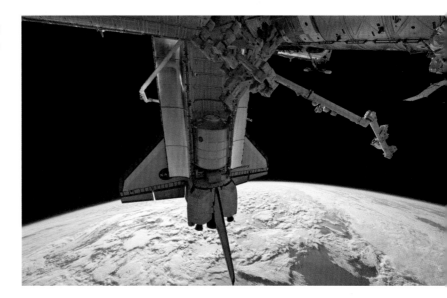

Lightweight Multi-Purpose Experiment Carrier (LMC)

The LMC is attached in the Shuttle cargo bay to carry equipment, supplies, or ORUs up to the ISS, and is moved off to the station using robotic manipulators. On this flight, the 1,100lb (500kg) LMC carried the 3,890lb (1,764kg) ammonia tank and returned with the old one, weighing 3,740lb (1,696kg), removed from the Shuttle.

STS-132/ISS ULF4/ MRM-1 Rassvet

Atlantis
14 May 2010, 6.20pm GMT
Cdr: Kenneth Ham (1)
Pilot: Dominic A. Antonelli (1)
MS1: Garrett Reisman (1)
MS2: Michael T. Good (1)
MS3: Stephen G. Bowen (1)
MS4: Piers J. Sellers (2)

The penultimate flight by *Atlantis* consisted of delivering the MRM-1 docking and cargo module, transferring the contents of a lightweight cargo carrier (VLD2), and supporting three spacewalks to carry out a wide range of repair, maintenance, installation, and ORU replacement tasks. Rassvet was removed from the cargo bay and docked to the nadir port on Zarya at 12.20pm on 18 May. Three EVAs were conducted for a total duration of 21hr 20min during which Reisman, Bowen, and Good

installed a spare SAGANT Ku-band antenna, positioned the EOTP for Dextre, and conducted general servicing tasks around the exterior of the station. Docked for 7 days 1hr 1min, *Atlantis* separated from the ISS and landed at the Kennedy Space Center on 26 May at 12.49pm GMT for a mission time of 11 days 18hr 29min.

Anatomy

Mini-Research Module-1 (MRM-1 Rassvet)

Length	19.7ft (6m)
Diameter (external)	7.7ft (2.3m)
Weight	11,188lb (5,075kg)
Weight (loaded at launch)	17,760lb (8,056kg)
Pressurised volume	614ft^3 (17.4m^3)
Habitable volume	207ft^3 (5.9m^3)

MRM-1 is a docking and cargo module designed and built by RSC Energia and attached to the nadir (Earth-facing) port on Zarya. The background to MRM-1 is a complex story framed by the changing financial situation in Russia and evolving requirements on the ISS. Russia initially planned to launch a docking and stowage module about the size of Zarya on a Proton rocket for docking to Zarya's nadir port, but that was cancelled due to economic pressure and the requirement down-sized to MRM-1. With a nominal crew of six, the ISS needed four docking ports capable of accepting Soyuz and Progress spacecraft, and with the permanent docking of Leonardo, the Russian segment would have only three. In addition, because the partnership deal had NASA launching equipment for the permanently placed Leonardo (from STS-133), it could also carry MRM-1.

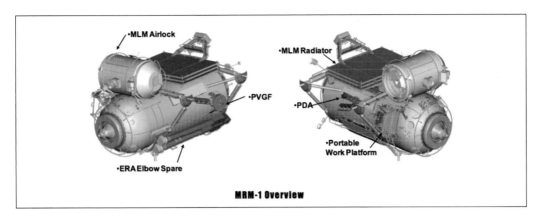

MLM Airlock

MLM Radiator

PVGF

PDA

Portable
Work Platform

ERA Elbow Spare

MRM-1 Overview

LEFT The Rassvet
docking module was
manufactured by RSC
Energia in Russia
and has evolved
from a chequered
background. (NASA)

Rassvet comprises a cylindrical module and has a docking port at each end. One is the docking interface with the nadir port on Zarya and the other can accept either Soyuz or Progress spacecraft, making this the fourth such docking facility on the Russian segment. The module also supports full environmental conditioning equipment and includes a flat radiator on the exterior surface. An experiment airlock on the side of the module provides for a variety of equipment.

During its flight to the ISS, Rassvet carried a total of 6,482lb (2,940kg) of cargo on internal and external stowage locations, and on the exterior of the cylindrical module it carried a spare elbow joint for the European Robotic Arm. It also carried equipment that would support the Russian Multi-Purpose Laboratory Module, tentatively scheduled for launch to the station in late 2012.

Anatomy

ICC-VLD2

A variation on the standard Integrated Cargo Carrier, this version had the suffix Vertical Light Deploy (VLD2) to indicate that it could be removed from the Shuttle's cargo bay by the RMS. The first VLD had been launched on STS-127 (see payload description) and VLD2 similarly carried six battery ORUs for the P6 array and one SAGANT communication array. VLD2 also carried an Enhanced Orbital Replacement Unit Temporary Platform (EOTP) which, along with the SAGANT equipment, would be located on ESP-3 at the P3 truss segment. Provided by Canada, the EOTP supported operation of the Special Purpose Dextrous Manipulator (Dextre) and was to be left at the station.

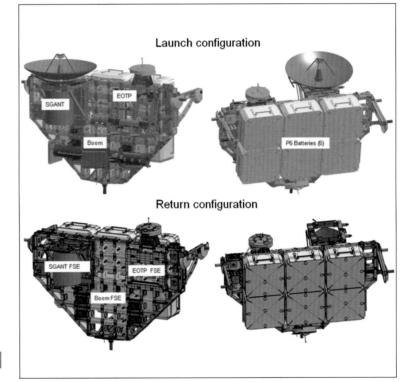

Launch configuration

SGANT

EOTP

Boom

P6 Batteries (6)

Return configuration

SGANT FSE

EOTP FSE

Boom FSE

Soyuz TMA-19/ISS 23S/ Expedition 24/25

15 June 2010, 9.35pm GMT
Cdr: Fyodor Yurchikhin (Russia)
Flight Engineer: Shannon Walker
Flight Engineer: Douglas H. Wheelock

Docking to the aft Zvezda port at 10.25pm GMT on 17 June, the crew joined Skvortsov, Korniyenko, and Caldwell Dyson – balancing the Expedition 24 team at three Russians and three Americans – until the resident crew returned on 25 September. The TMA-19 crew returned to their spacecraft on

ABOVE Up and down loads for the ICC-VLD equipment and supplementary cargo carrier aboard *Atlantis* on STS-132. (NASA)

28 June to relocate it at the Rassvet module, vacating the aft Zvezda port for the arrival of Progress M-06M on 4 July. Yurchikhin, Walker, and Wheelock received the crew of TMA-01M on 9 October and briefly shared duties and tasks until returning to Earth the following month.

TMA-19 undocked from the ISS at 1.19am on 26 November and landed at 4.46am the same day, four days earlier than originally planned to avoid air traffic restrictions by the Kazakh authorities in preparation for a European security summit held in Astana on 1–2 December. Recovery forces would have been grounded and unable to reach the spacecraft had the landing held to its planned date of 29 November.

Soyuz TMA-01M/ISS 24S/ Expedition 25/26

7 October 2010, 11.11pm GMT
Cdr: Aleksandr Kaleri (Russia)
Flight Engineer: Oleg Skripochka (Russia)
Flight Engineer: Scott Kelly

Marking the first launch of an improved and upgraded version of the Soyuz spacecraft, TMA-01M docked with the Poisk module at a minute past midnight GMT on 10 October, carrying the Expedition 25 crew, joining Yurchikhin, Walker, and Wheelock for almost seven weeks before they returned to Earth on 26 November. The Expedition 25 crew remained aboard the ISS for more than six months, receiving during their stay the crew of TMA-20, the second of Japan's Kounotori logistics modules, the second of ESA's Automated Transfer Vehicles, and hosting the

last flight of *Discovery* (STS-133). TMA-01M undocked at 4.27am GMT on 16 March 2011, and returned to a landing on the snow-covered region of Arkalyk under high winds and sub-zero temperatures precisely 3hr 30min later.

Soyuz TMA-20/ISS 25S/ Expedition 26/27

15 December 2010, 7.09pm GMT
Cdr: Dmitri Kondratyev (Russia)
Flight Engineer: Catherine Coleman
Flight Engineer: Paolo A. Nespoli (ESA)

Before the flight of TMA-20 engineers noticed damage to the descent module, which was replaced with the module originally built for TMA-21, but the launch went well and the crew docked with the Rassvet module at 8.12pm GMT on 17 December.

TMA-20 undocked from the ISS at 2.35pm GMT on 23 May 2011 as the assembly of orbiting space vehicles was passing over eastern China. Uniquely, representative examples of all the primary modules and spacecraft that had participated in the assembly of the station over many years were present at the ISS. These included Russia's Progress and Soyuz vehicles, NASA's Shuttle, Japan's Kounotori 2, and Europe's ATV-2 and the MPLM Leonardo, now renamed the Permanent Multipurpose Module. By agreement with the Russians, NASA masterminded a photo shoot of the complete assembly, conducted by Nespoli as he viewed the station from Soyuz TMA-20.

With Kondratyev at the controls from his centre-couch position, the spacecraft backed away 600ft (183m) while to his left Nespoli took still photographs and video. Opening the hatch to the orbital module, Nespoli took additional views from a porthole. Just 30 minutes after undocking, the station began to gently rotate by 129 degrees in order to provide the best illuminated conditions, thereby adding a further 25 minutes to the photo shoot. With Nespoli back in the descent module the separate sections of the Soyuz were jettisoned and TMA-20 landed back on Earth at 2.27am GMT on 24 May 2011, having spent 159 days 7hr 17min in space.

HTV-2/ISS-HTV2 Kounotori 2

22 January 2011, 5.38am GMT

A repeat mission of that flown by HTV-1 and following procedures pioneered on that mission, the second of Japan's expendable logistics carriers docked to the Harmony nadir port at 2.51pm GMT on 27 January. On board were 11,686lb (5,300kg) of stores consisting of 8,820lb (4,000kg) in the Pressurised Logistics Carrier (PLC) and the remainder in the Unpressurised Logistics Carrier (UPLC).

Just over half the weight of freight in the PLC comprised spare components. The rest was food, science equipment, crew items, and water. The load included the Kobairo gradient heating furnace rack to be used for heating up large-scale high-quality crystals, and a small multipurpose payload both of which were transferred into the Kibo research module after docking. The UPLC carried an Exposed Pallet stashed with two ORUs, a flexible hose rotary coupler, and a cargo transportation container transferred to the external platform using Kibo's remote manipulator.

Timed operations with HTV-2 were compromised by the late arrival of STS-133, originally scheduled for September 2010 and carrying a Logistics Carrier to which the external payload would be attached. HTV-2 would also receive waste from the Shuttle so it had to remain attached to the Harmony module longer than planned. After sealing it back up for undocking, on 18 February HTV-2 was relocated to Harmony's zenith port using Canadarm2.

After *Discovery* departed on 7 March, Kounotori 2 was relocated back to the zenith port three days later but when a catastrophic Earthquake hit Japan the following day it damaged the Tsukuba control centre and its departure was delayed further until resumption of operations on 22 March. Six days later, at 3.29pm GMT, Canadarm2 removed it from the nadir port and 15 minutes later released it for atmospheric re-entry.

Anatomy

H-II Transfer Vehicle

For details see HTV-1 payload description, 10 September 2009.

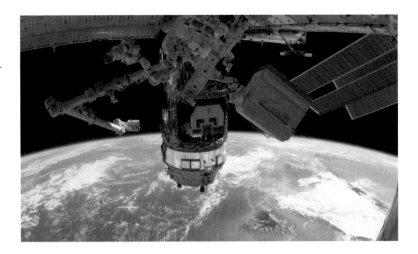

ABOVE With Dextre in the foreground, HTV-2 is firmly berthed to the nadir port on Harmony. *(NASA)*

LEFT Japan's HTV-2 logistics carrier arrives at the ISS with more than 5 tonnes of supplies. *(JAXA)*

ATV Johannes Kepler/ ISS-ATV-2

16 February 2011, 9.51pm GMT

The second of ESA's Automated Transfer
Vehicles, Johannes Kepler was launched
by Ariane V with cargo, freight, supplies, and
equipment primarily for the Expedition 26 crew
and for the visit of STS-133. With on-board wet
and dry supplies it weighed 44,100lb (20,000kg)
at launch and docked to the aft Zvezda port
at 3.59pm GMT on 24 February, shortly before
the launch of STS-133. Details of the ATV can
be found in the description of the Jules Verne
module launched on 9 March 2008.

ATV-2 carried 15,620lb (7,085kg) of cargo
including 10,000lb (4,536kg) of ISS propellants,
1,900lb (862kg) for refuelling Zvezda's

propulsion unit, 220lb (100kg) of oxygen, and
3,500lb (1,587kg) of dry cargo including food,
clothes, and equipment. Kepler's propulsion
system was used on 18 March to re-boost the
ISS, increasing altitude by approximately 2nm.
The mission ended when ATV-2 separated
from the ISS on 20 June 2011, returning
to destruction in the Earth's atmosphere at
10.44pm GMT.

STS-133/ISS ULF5/ Robonaut 2

Discovery
24 February 2011, 9.53pm GMT
Cdr: Steven W. Lindsey (4)
Pilot: Eric A. Boe (1)
MS1: Alvin Drew (1)
MS2: Steve Bowen (2)
MS3: Michael Barratt
MS4: Nicole Stott (1)

A logistics and delivery mission, STS-133
delivered the Leonardo MPLM which,
when docked to the station, would be renamed
the Permanent Multipurpose Module, never to
return to Earth. It would also deliver ExPRESS
Logistics Carrier 4 (ELC-4) and place the first
humanoid robot – Robonaut – aboard the
station. ELC-4 would be positioned on the
lower inboard position of the starboard side
of the station's truss structure and Robonaut
2 would be unpacked and stored aboard the
station for later use. Of the crew, Barratt was
the only astronaut on the flight not to have flown
a Shuttle mission before, his experience having
been on Soyuz TMA-14. He replaced Tim
Kopra, who was injured in a bicycle accident a
month before lift-off.

Docking with the ISS at 7.14pm GMT on
26 February, the hatches were opened two
hours later and the crew of STS-133 joined the
Expedition 25 team for nine days of intense
activity. The Leonardo module was removed
from the cargo bay and attached to the ISS on
1 March and two days later the crew began to
unload it. Two spacewalks were conducted.
The first, lasting 6hr 34min, allowed Bowen and
Drew to install a new power interlink between
the Unity and Tranquility modules and move a
failed ammonia pump from its location on the

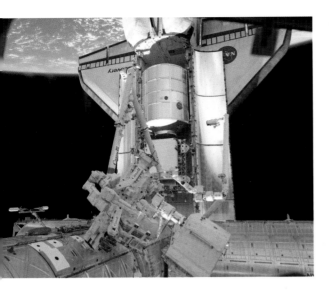

ABOVE A view through the windows on the Cupola show *Discovery* with the Leonardo module and ELC-4 still aboard, Canada's comprehensive robotic suite of support equipment clearly visible in the foreground. *(NASA)*

ESP-2 to the Quest airlock module. Two days later the second EVA lasted 6hr 14min and completed servicing and maintenance tasks aboard the ISS.

Undocking from the ISS at 12.01pm GMT on 7 March, *Discovery* returned to Earth for the last time, making its 39th landing back from space at 4.58pm on 9 March, ending a flight that lasted 12 days 19hr 5min, the most flown Orbiter in the fleet.

Anatomy

Robonaut 2

Development of a humanoid robot that could provide an extra pair of hands for complex tasks or work outside the station on lengthy but mundane tasks began in 1997 with Robonaut 1, designed to be capable of exploring the Moon or Mars. When funding ran out in 2006, General Motors took up the challenge and in exchange for NASA know-how provided the money to develop what became Robonaut 2. Adapted for working at the ISS, Robonaut 2 was lifted to the station packed inside the Leonardo logistics module and placed aboard the station, where over the next several months it would be deployed and tested in a variety of experimental procedures.

Precisely because it is designed to mimic the operating dexterity of a human, it necessarily comes out looking humanoid, but its head contains its sensory systems and not its 'brain'. Behind its 'face' are four cameras working in the visible portion of the spectrum – two for stereo vision – and a fifth infrared camera in the 'mouth' area. Its 'neck' has three degrees of freedom allowing it to look left, right, up, or down, and each arm has seven degrees of freedom, enabling it to hold a 20lb (9kg) mass in any pose.

Each arm section (upper and lower) has a length of 2ft 8in (81cm) giving a reach of 8ft (2.4m). It is said that the single most distinguishing feature of humans among all primates is the dexterity of the hand, so each hand on Robonaut 2 has 12 degrees of freedom: four in the thumb, three each in

ABOVE A development from General Motors, Robonaut 2 was carried to the ISS by *Discovery* for extensive tests in man–machine cooperation inside the pressurised laboratory. *(NASA)*

LEFT Dan Burbank gets to know Robonaut 2 as the astronauts test the ability of robotic machines to perform low-level tasks on the ISS. *(NASA)*

the index and middle fingers, and one each in the ring and pinky fingers. Each finger has a grasping force of 5lb (2.3kg). Power for Robonaut 2 comes from an electrical cable from its back-mounted power conversion system to the ISS supply, but its torso contains the electronic processing system to control its functions.

Made of nickel-plated carbon fibre and aluminium, Robonaut 2 has a height of 3ft 4in (102cm) from waist to head, a shoulder width of 2ft 7in (79cm) and weighs about 300lb (136kg). It has 42 sensors for degrees of freedom and 38 PC processors operating off 120VDC supplied by the station's electrical system. Before going into space Robonaut 2 had to prove it was resistant to vibrations, did not impose unacceptable operating noises, did not emit electromagnetic waves that could interfere with equipment, and was capable of being unpacked in weightlessness.

Anatomy

Permanent Logistics Module

Length	21ft (6.4m)
Diameter	15ft (4.6m)
Weight at launch	28,353lb (12,861kg)
Weight empty	21,817lb (9,896kg)

Manufactured by Thales Alenia Space in Italy and flown seven times as the Leonardo Multi-Purpose Logistics Module (MPLM), for this eighth flight the carrier was rebranded as the Permanent Logistics Module (PLM) to adopt a resident role at the station, attached to the nadir port on Unity (Node 1). On its final journey to the station, the PLM carried 14 racks: one experiment rack, six Resupply Stowage Platforms, five Resupply Stowage Racks, and two Integrated Stowage Platforms. The experiment rack consisted of ExPRESS rack 8 supporting science experiments in a multitude of disciplines. In addition, several spare assemblies were lifted to the ISS inside the PLM, including a common air assembly heat exchanger for controlling temperature and humidity, a device of 34.8in x 21.4in x 17.5in (88cm x 54cm x 44cm) in size and weighing 125lb (57kg).

Another spare was the pump package assembly as a second back-up unit for circulating coolant fluid throughout the internal thermal control system, measuring 29.6in x 18.6in x 17.75in (75cm x 47cm x 45cm) and weighing 191lb (87kg). Simply known as the Inlet, a large 59lb (27kg) cabin fan measuring 17.5in x 21in by 24in (44cm x 53cm x 61cm) was lifted to the ISS to take in the station atmosphere and circulate it for mixing. A spare water processor assembly water storage

RIGHT The Leonardo MPLM is made ready for its final flight to the ISS where it was left permanently docked at the nadir (Earth facing) port on Unity. *(NASA)*

tank was included in the manifest. Weighing 154lb (70kg), the 34.74in x 17.23in x 19.38in (88.2cm x 43.8cm x 49.2cm) device was manufactured by Hamilton Sundstrand and consists of a bellows tank with quantity sensors, solenoid valves, and quality sensors for determining the chemical characteristics of the water. An associated waste water tank, 32.54in x 17.3in x 18.77in (82.7cm x 43.9cm x 47.7cm) and weighing 230lb (104kg), holds 100lb (45kg) of condensate from the urine and waste water system.

Anatomy

ExPRESS Logistics Carrier 4 (ELC-4)

Similar to ELC-1 and -2 carried to the ISS aboard STS-129 (see that mission's payload description), ELC-4 was positioned on the S3 truss segment lower inboard Passive Attach System (PAS). It supported just one payload, the Heat Rejection Subsystem Radiator (HRS) comprising a base plate, torque arm, eight panels, an interconnected fluid system, and a computer-controlled deployment mechanism. The HRS is part of the active thermal control system designed to discharge heat from the station's operating systems via radiation to the vacuum of space. The HRS radiator unit carried by STS-133 is a spare for one of the six that comprise part of the External Active Thermal Control System (EATCS), three each on S1 and P1. Each radiator weighs 2,475lb (1,123kg) and has a length of 74.6ft (22.7m), a diameter of 11.2ft (3.4m), and a surface area of 1,554ft^2 (144m^2). The total weight of ELC-4 was 8,235lb (3,735kg).

Soyuz TMA-21/ISS 26S/ Expedition 27/28

4 April 2011, 10.18pm GMT

Cdr: Aleksandr Samokutyayev (Russia)
Flight Engineer: Andrei Borisenko (Russia)
Flight Engineer: Ronald J. Garan (US)

The three-man crew of TMA-21 docked with the Poisk module at 11.29pm GMT on 6 April, the hatches opening at 2.13am GMT the morning after, restoring the crew complement aboard the ISS to six. STS-134 arrived on 17 May and for several days there were 12 people at the station. The resident crew of Kondratyev, Coleman, and Nespoli returned to Earth on 24 May.

They received a further complement to the Expedition 28 group when Soyuz TMA-02M arrived in June and they hosted the last Shuttle visit when *Atlantis* made its last flight to the ISS in July. The TMA-21 crew had expected to return to Earth on 8 September but the loss of Progress M-12M on 24 August when the upper stage of its launch vehicle failed delayed their return. TMA-21 finally undocked from the ISS at 12.38am GMT on 16 September and landed at 4.00am GMT that day.

STS-134/ISS ULF6

Endeavour
16 May 2011, 12.56pm GMT

Cdr: Mark E. Kelly (3)
Pilot: Gregory H. Johnson (1)
MS1: Michael Fincke
MS2: Greg Chamitoff (1)
MS3: Andrew Feustel (1)
MS4: Roberto Vittori (ESA)

Before the flight of STS-134 serious doubts were cast on the likelihood of Mark Kelly commanding this mission. His wife, Congresswomen Gabrielle Giffords, had survived

ABOVE The ELC-4 logistics pallet is packed for flight on STS-133. *(NASA)*

ABOVE ESA astronaut Roberto Vittori floats through the US Destiny module. *(NASA)*

BELOW With Dextre getting an arm extension from the station's remote manipulator, ELC-3 is slowly lifted into position at the P3 truss segment. *(NASA)*

an assassination attempt on 8 January 2011, and was recovering from a near-fatal injury. Mark's twin brother Scott had flown in space as a crewmember of Expedition 26, making them the only twins and the only siblings to have flown into space. On 14 March sadness enveloped the launch preparation teams at the Kennedy Space Center when an engineer, James Vanover, committed suicide by jumping from the launch gantry where *Endeavour* was being made ready for flight. But when *Endeavour* was launched on its last mission, Gabrielle Giffords made a brave journey to witness the event.

Following a flawless launch, docking with PMA-2 occurred at 10.14am GMT on 18 May and the six crewmembers quickly joined the Expedition 27 crew aboard the ISS,

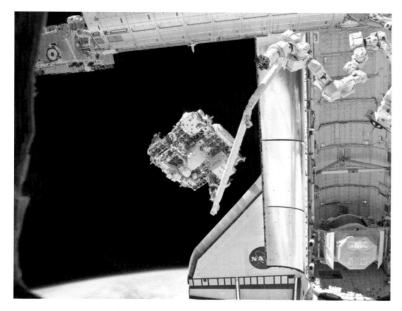

hatches opening at 11.30am GMT. With a very busy flight plan the crew got to work, using Canadarm2 to lift the ELC-3 carrier from the payload bay and locate it at the P3 truss segment. The Alpha Magnetic Spectrometer was lifted from the Shuttle on 19 May, with final installation on the S3 truss segment being completed at 9.46am GMT. Four spacewalks totalling 28hr 44min saw, at various times on 20, 22, 25, and 27 May, Feustel, Chamitoff, and Fincke carry out a wide range of installation and servicing tasks as well as changing out the MISSE materials science experiments on the exterior of the station.

In a serious attempt to understand the challenges of working with small pieces of an assembly kit, the Shuttle carried 12 Lego packs for the astronauts to build, evaluating difficulties or problems and assessing the ease with which they were able to complete the models. Not a few thought this was a very good excuse to have some fun! Left behind on the ISS was *Endeavour*'s Orbiter Boom Sensor System (OBSS) to extend the reach of Canadarm2, now no longer needed by *Endeavour*. The crew also conducted several experiments in space physics for the Air Force in its Space Test Program.

As recorded earlier, when Soyuz TMA-20 separated on 24 May, ESA astronaut Paolo Nespoli took a sequence of stunning images showing the complete array of spacecraft types that had visited the station in the lengthy assembly period that began in 1998. Less than a week later it was also time for *Endeavour* to return to Earth. Undocking came at 3.55am GMT on 30 May, some 11 days 17hr 41min after it latched on to the ISS. The Shuttle landed on 1 June at 6.35am GMT after a flight duration of 15 days, 17 hours and 39 minutes.

Anatomy

Alpha Magnetic Spectrometer 2 (AMS-2)

Weight	14,809lb (6,717kg)
Power demand	2–2.5kW
Internal data rate	2Mbits/sec
Mission lifetime	10–18 years

The AMS is a particle physics experiment designed to search for unusual matter in the universe by measuring cosmic rays, energetic charged magnetic particles originating in some of the most dynamic events in space. It is through the study of these particles that physicists hope to explain dark matter and anti-matter. The origin of the AMS goes back to the cancellation of the Superconducting Super Collider in October 1993.

Less than two years later Nobel laureate and MIT physicist Samuel Ting proposed an Alpha Magnetic Spectrometer attached to the space station, by that time the beneficiary of a truly international consortium including the Russians. A prototype, AMS-1, was proposed by Prof Ting and this was built and flown aboard *Discovery* on STS-91 when it made the last Shuttle flight to the Mir station in June 1998. It was a success and proved Prof Ting's concept.

On the basis of these results, Prof Ting gathered a working group of 500 scientists from 16 countries in a programme funded by the US Department of Energy. An original cost estimate of $33m increased to a final cost of $1.5bn after design and manufacturing difficulties and delays brought about by the loss of *Columbia* in 2003, the year in which it was originally to have flown.

There was a time when it was thought AMS-2 would never fly, the Shuttle programme being pressed by some to retire after the STS-133 mission, but Congress approved two more missions and the project

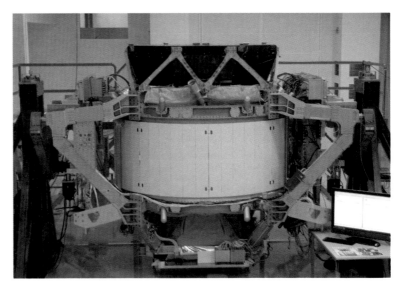

was saved. As installed, it contains two coils of niobium-titanium, producing a central field of 0.87 teslas in a non-superconducting system for increased life.

ABOVE Precursor to the advanced astrophysical experiment lifted to the ISS, the first Alpha Magnetic Spectrometer (AMS-1) was tested in space aboard *Discovery* in 1998. *(NASA)*

LEFT AMS-2 is packaged aboard *Endeavour* for installation on the ISS in May 2011. *(NASA)*

LEFT In place on the S3 truss segment, AMS-2 began sending back highly valuable scientific data from the first day it went into operation. *(NASA)*

ExPRESS Logistics Carrier 3 (ELC-3)

The fourth and last ELC structure lifted to the ISS carried a variety of experiment packages and support equipment and was attached to the P3 truss segment Passive Attach System (PAS). It weighed 14,023lb (6,361kg). Previously, ELC-1 and -2 had been lifted to the ISS on STS-129, followed by ELC-4 on STS-133 (see those mission payload descriptions).

ELC-3 carried one pack of equipment designed to supply back-up Orbital Replacement Units (ORUs) to equipment and payloads brought up on earlier flights. One of those was a Cargo Transportation Container (CTC), to which were attached ten Remote Power Controller Modules and 11 ORU adapter kits to hold them in the CTC. The container also had an Arm Computer Unit ORU, a vital part of Canadarm2 for which an on-site replacement was essential, bringing a high level of redundancy to this robotic system. The CTC weighed 1,050lb (476kg) for this mission, somewhat short of its maximum capacity.

Several other ORUs were aboard ELC-3, including an Ammonia Tank Assembly (ATA) in which ammonia coolant fluid is stored for the External Thermal Control System (ETCS), the complete package including two storage tanks, isolation valves, heaters, and quantity/quality sensors. One ATAS loop is attached to each loop, one on the zenith side of the Starboard 1 (loop A) truss segment and another on the Port 1 (loop B) segment. The package is 4ft 9in long by 6ft 8in wide and 3ft 9in tall (1.4m x 2m x 1.1m), weighing when new a total of 1,702lb (772kg) including 600lb (272kg) of ammonia.

Carried to the station on ELC-3 was a High-Pressure Gas Tank (HPGT), a vital function of which was to supply oxygen from a stowed position within the Quest airlock module, to which it was transferred. The HPGT weighs 1,240lb (562kg), of which 220lb (100kg) is gaseous oxygen stored at 2,450psi, in a tank that measures 5ft x 6.2ft x 4.5ft (1.5m x 1.9m x 1.4m).

The S-Band Antenna Support Assembly (SASA) contains the radio frequency equipment for S-band communications with the Tracking and Data Relay Satellite. It consists of a mast, an EVA handle, harness, connector panel, mounting surfaces, and a base-plate fitting. The total SASA envelope is 36in x 59in x 33in (91cm x 150cm x 84cm) and the boom itself is 61in x 30.25in x 43in (155cm x 77cm x 109cm) with a weight of 256lb (116kg). ELC-3 provided two spare SASA units in addition to the on-orbit spare delivered aboard STS-129. Another item delivered to the station was a spare arm for Dextre (Special Purpose Dextrous Manipulator), 11.5ft (3.5m) long incorporating seven joints and with a load carrying capacity of 1,320lb (599kg).

Materials on International Space Station 8 (MISSE-8)

MISSE-8 was carried aboard the Shuttle latched to the port side of the cargo bay and emplaced by the robotic arm in return for MISSE-7, which had been at the ISS since delivered in November 2009. MISSE-1 and -2 had been launched by STS-105 in August 2001 and recovered by STS-114 four years later in return for leaving MISSE-5, which was retrieved by STS-115 in September 2006. MISSE-3 and -4 were delivered by STS-121 in July 2006, returning aboard STS-118 in August 2007. MISSE-6A and -6B were carried to the ISS by STS-128 in September 2009, followed by MISSE-7 with STS-129 two months later.

Soyuz TMA-02M/ISS 27S/ Expedition 28/29

7 June 2011, 8.12pm GMT
Cdr: Sergey Volkov (Russia)
Flight Engineer: Satoshi Furukawa (Japan)
Flight Engineer: Michael E. Fossum (US)

This was a routine crew delivery flight taking the Expedition 28 crew to join the host team of Samokutyayev, Borisenko, and Garan, who had been on board the ISS for two months. Docking with the Rassvet module came at 5.18pm GMT on 9 June. The host crew returned to Earth in their Soyuz TMA-21 spacecraft on 16 September, when the mission became Expedition 29, but by this date the Russian-manned flight programme had been grounded.

A failure in the upper stage of a Soyuz launch vehicle carrying Progress M-12M on 24 August brought a halt to launch operations until the problem had been understood and any modifications made. This was an identical stage to that used to carry the manned Soyuz spacecraft and it could not be cleared for flight until the problem had been fully understood. Moreover, having retired its Shuttle programme with the landing of *Atlantis* on 21 July, there was now no manned vehicle capable of reaching the ISS.

While the three remaining crewmembers of Expedition 29 could return at any time if any danger should present itself, there was a limit on the time their Soyuz TMA-02 spacecraft could remain in space – 220 days. If the next Soyuz could not be launched before mid-December the TMA-02M crew would have to vacate the station and return, leaving it to be operated without a crew for the first time in more than a decade.

The Russians worked quickly, and the failure was assessed and found to have no impact on a manned launch. The flight of TMA-22 was rescheduled for 14 November 2011, with Volkov, Furukuwa, and Fossum returning to Earth in TMA-02M at 2.26am GMT on 22 November.

STS-135/ISS ULF7

Atlantis
8 July 2011, 3.29pm GMT
Cdr: Chris Ferguson (2)
Pilot: Doug Hurley (1)
MS1: Sandra Magnus (2)
MS2: Rex Walheim (2)

Inserted as a bonus logistics and delivery flight, STS-135 was planned as a mission to stock up the ISS with a wide range of spares, supplies, and cargo to operate the station over the period between the last flight of the Shuttle and the first commercial logistics delivery flights planned for 2012. Veteran astronaut Sandra Magnus was loadmaster and with her experience of the station from previous flights she was charged with unpacking, and packing up, the Raffaello MPLM and transferring equipment off the Lightweight Multi-Purpose Carrier. No spacewalks from the Shuttle crew were to be performed, although station-based EVAs were conducted. Partly because of this and to eliminate unnecessary congestion

LEFT The ISS Expedition 28 crew and the four crewmembers of STS-135 get together for a photo-call. *(NASA)*

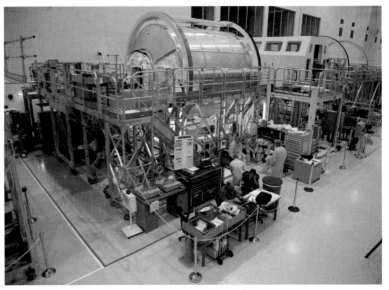

during the busy work of moving bulky cargo, the four-person crew was the smallest of any since STS-6 in 1983.

After two days of flight to the ISS, docking occurred at 3.07pm GMT on 10 July, followed by the removal of Raffaello from the cargo bay to a docking at the nadir port on Node 2 at 10.46am GMT on 11 July. Raffaello carried its maximum load of eight Resupply Stowage

ABOVE Rafaello is packed up ready to fly to the ISS aboard *Atlantis* **in July 2011 on the final Shuttle mission.** *(NASA)*

LEFT The US Honor Flag that flew for two weeks at ground Zero after the devastating events of 9/11 is now aboard the International Space Station where it will remain until returned to Earth by the next US crew flying to the ISS in a spacecraft made in the USA. *(NASA)*

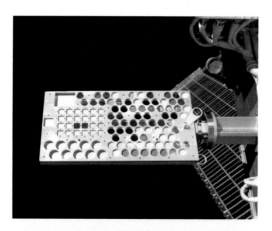

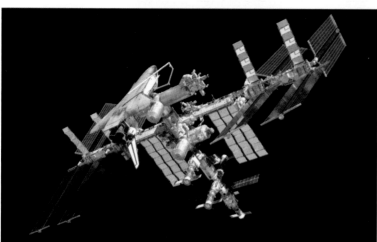

Platforms, two International Standard Payload
Racks, six Resupply Stowage Racks and one
Zero-G Stowage Rack. This was the fourth flight
of Raffaello, which for this mission was stocked
with 9,403lb (4,265kg) of freight for the ISS,
receiving 5,666lb (2,570kg) for return to Earth.
In addition, 1,564lb (709kg) of cargo had been
stocked in the Shuttle middeck area and after
this had been transferred to the ISS, 1,564lb
(709kg) was loaded in from the station.

The crew had their work cut out to
accomplish this in the time available and Sandra
Magnus was officially in charge of the operation,
her knowledge of the ISS playing an important
role in facilitating efficient transfer and stowage
of all the bags, bundles, and freight moved
back and forth. Behind Raffaello in the Shuttle's
cargo bay the seventh Lightweight Multi-
Purpose Carrier (LMC) was installed, a 946lb
(429kg) Shuttle cargo bridge holding 2,918lb
(1,324kg) of equipment for transfer to the ISS
including the Robot Refueling Mission. The
LMC served as a mounting for a returned Pump
Module which had failed on 31 July 2010, and
was replaced two weeks later.

Following seven days of intense labour, and
carrying 3,530lb (1,600kg) of cargo offloaded
from the ISS, Raffaello was unberthed from
Node 2 at 10.48am GMT on 18 July. It was
back in *Atlantis* one hour later, completing the
12th flight of an MPLM since the first, Leonardo,

FAR LEFT The launch of Soyuz TMA-03M in December 2011 brought three crewmembers to the ISS, restoring the complement to six people. *(NASA)*

LEFT ESA astronaut Gerard Kuipers (top) flew with NASA astronaut Don Pettit and Russian cosmonaut Oleg Kononenko. *(NASA)*

had been carried to the station in March 2001. The second MPLM, Raffaello, had made four trips. After 8 days 15hr 21min attached to the ISS, *Atlantis* undocked from PMA-2 at 6.28am GMT on 19 July. On behalf of the entire Shuttle programme it said farewell for the last time, leaving the six crewmembers of Expedition 28 at the ISS. *Atlantis* returned to Earth at 9.57am GMT, just before dawn on 21 July, after a flight lasting 12 days 18hr 29min.

Anatomy
Robotic Refueling Mission (RRM)

Length	3ft 9in (114cm)
Width	2ft 9in (84cm)
Height	3ft 11in (119cm)
Weight	550lb (250kg)

The LMC carried to the station a device for testing the possibility of one day servicing satellites in space without human intervention. Known as the Robotic Refueling Mission (RRM), it was a joint project between NASA and the Canadian Space Agency to test the technical feasibility of refuelling satellites not designed for such a task. It would also require the use of Dextre for the first time on jobs other than maintenance on the station, applying it to a productive experiment. The RRM was moved to the Enhanced Orbital Replacement Unit Temporary Platform (EOTP) to which Dextre was attached and, after the departure of the Shuttle, Canadarm2 relocated it to its permanent site on ELC-4. Remotely controlled by NASA engineers on Earth, the RRM would evaluate the ability of the robot to use a suite of tools to access a satellite and insert a fuel valve and refuel it for further use.

Soyuz TMA-22/ISS 28S/ Expedition 30

14 November 2011, 4.14am GMT
Cdr: Anton Shkaplerov (Russia)
Flight Engineer: Anatoli Ivanishin (Russia)
Flight Engineer: Daniel C. Burbank (US)

This was the last flight of the Soyuz TMA vehicle, future flights being with the TMA-M series. Docking occurred at 5.24am GMT on 16 November over the Pacific Ocean and the crew entered the ISS to join Fossum, Volkov, and Furukawa about 75 minutes later. For five days the six crewmembers conducted a handover before the resident crew separated on 21 November and returned to Earth, reducing the ISS to a complement of three until they were joined by the crew of TMA-03M, launched on 21 December.

Soyuz TMA-03M/ISS 29S/ Expedition 30/31

21 December 2011, 1.16am GMT
Cdr: Oleg Kononenko (Russia)
Flight Engineer: Andre Kuipers (ESA)
Flight Engineer: Donald Pettit (US)

Coming only six weeks after the launch of TMA-22, the flight of TMA-03M restored the ISS to its complement of six personnel, comprising three cosmonauts, two Americans, and a European. When launched the mission was expected to host visits by several unmanned vehicles: Europe's ATV-3 (Edoardo Amaldi), two US commercial logistics supply modules (SpaceX Dragon and Orbital Sciences Cygnus module), and Russian Progress cargo-tankers.

RIGHT Research aboard the ISS is focused around a set of experiment cabinets common to all modules, as seen here for the US Destiny laboratory. (NASA)

RIGHT Research aboard the ISS is focused around a set of experiment cabinets common to all modules, as seen here for the US Destiny laboratory. (NASA)

FAR RIGHT Expedition 30 commander Dan Burbank works in Destiny with equipment designed to study high magnification colloid experiments. (NASA)

BELOW ESA's microgravity science glovebox for handling contained experiments. (ESA)

U.S. Laboratory Module
International Space Station

Materials Sciences
Standard Rack-1
(example)

EXPERIMENTS ABOARD THE ISS

Key to effective scientific operation of the ISS was the International Standard Payload Rack (ISPR) which would be the standard experiment casing for all US, European, and Japanese modules. Standardisation helps not only with the fittings inside all experiment modules but also simplifies the logistical challenges of lifting modules to the station so that a commonly agreed size and volume for each rack from whichever country can fit inside any module or cargo freighter.

Each ISPR is outfitted on the ground prior to launch with up to 1,540lb (700kg) of equipment and is conveyed to the ISS on a Multi-Purpose Logistics Module (MPLM). Each rack is 73in (1.85m) tall, 33.8in (86cm) deep and 42in (1.07m) wide, is built around a graphite composite shell, and weighs about 1,200lb (544kg). Each standard rack has an internal volume of 53ft^3

(1.5m^3). Tracks on the front of the racks allow mounting of laptop or tablet computers and cables fed through utility ports provide power and data connections. One major problem with the ISPR system is that their size prohibits any vehicle except the Shuttle, now retired, or the H-II Transfer Vehicle from carrying them into the station. Russian APAS or NDS docking systems are too small and only vehicles connecting to a 50-inch square Common Berthing Mechanism can afford the open area for passage.

A standardised rack system allows experiments to move up and down between Earth and the ISS, or to remain aboard the station for extended periods. Known as ExPRESS (Expedite the Processing of Experiments to the Space Station) this provides a standard interface between equipment and the utilities outlets available on the ISS. The first two

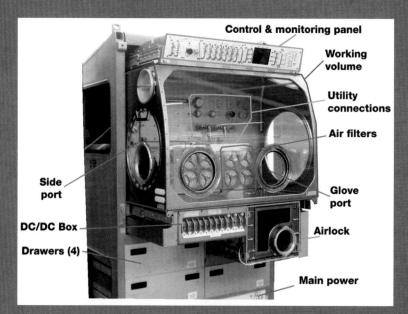

Control & monitoring panel

Working volume

Utility connections

Air filters

Side port

Glove port

DC/DC Box

Airlock

Drawers (4)

Main power

BELOW The ExPRESS rack is standard and allows scientists and technicians on the ground to prepare new or replacement experiments for uploading in logistics freighters. (NASA)

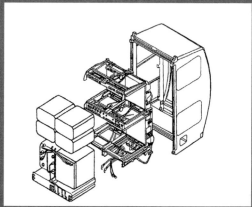

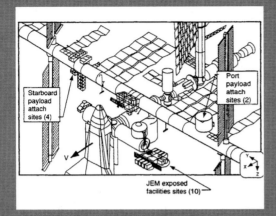

ExPRESS racks were carried aboard the STS-100 mission in April 2001 and were installed in an ISPR in orbit. These racks can be controlled by the crew on the station or by controllers on the ground and have been a direct result of tests conducted aboard independent Shuttle missions before assembly of the ISS began. The first such test was aboard STS-94 in 1997 and validated the sub-rack concept of breaking down ISPR shelves into carriers for separate experiments. This approach eases the availability of the ISS for colleges and universities to buy in to exactly the size and capacity of experiment volume they need for their work, and to pay for no more than they need.

BELOW Tools are vital for completing essential maintenance, repair and replacement activities outside the station, made easier with the use of this temporary worksite support. *(NASA)*

FAR LEFT Provision for payloads outside the ISS makes use of multiple locations on the truss segments and the exterior surfaces of pressurised modules. *(NASA)*

LEFT Cosmonaut Anton Shkaplerov conducts an EVA for routine activity outside the ISS during his visit with Expedition 30. *(NASA)*

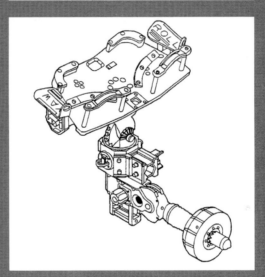

LEFT An articulating portable foot restraint, essential for maintaining the legs and lower torso in a fixed position during space walk activity. *(NASA)*

BELOW The tool board provides an adaptable fixture for a variety of commonly used tools and devices for working outside the station. *(NASA)*

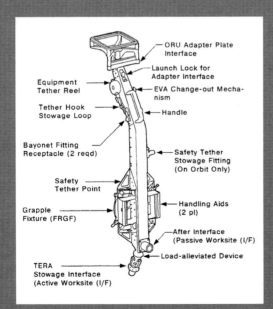

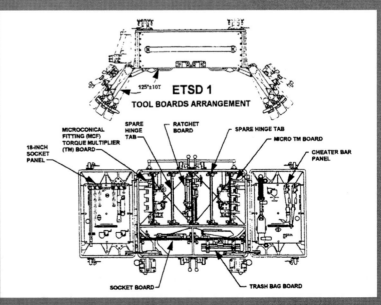

Chapter Six

Legacy

With assembly complete, the ISS was ready for a new phase of operations. The Shuttle was retired and commercial companies signed contracts to deliver cargo and people to the giant laboratory in space, a transformation that had never been envisaged when the station was approved almost 30 years earlier.

OPPOSITE Expedition 30 commander Dan Burbank wiles away free time playing his guitar in the Destiny module. *(NASA)*

RIGHT Although inside the Shuttle Middeck area, Karen Nyberg demonstrates the ability of weightlessness to convert long hair to a bouffant style! *(NASA)*

Completing NASA's 30-year long Shuttle programme, the last flight of *Atlantis* in July 2011 brought high emotion to the mission. It was for building and servicing a station in space that the Shuttle had been conceived, being adapted on the way to that objective into a space truck for launching satellites, sending spacecraft to the planets, placing astronomical observatories in orbit, and occasionally playing the role of a repair vehicle for stranded satellites. But it was the space station concept that in the mid-1960s drove NASA away from the one-shot spacecraft that preceded it to a reusable transportation system. Without that system the International Space Station as we know it today could never have been built.

The concept of a reusable Shuttle replaced all other ideas about how to increase reliability and use a small fleet of four Orbiters to launch, assemble, and service a station that would grow piece by piece, each section paid for and launched in instalments. It took a great global effort, one of the greatest gatherings of scientists and engineers since the International Geophysical Year of 1957–58, a period during which both Russia and the United States launched their first satellites. And among former ideological adversaries, the ISS bridged the period of Cold War confrontation and the new reality of peaceful cooperation in a world that has to live and work together to meet global challenges, be they man-made, environmental or economic.

Getting aboard the station

For a trip to the ISS, potential visitors need to have one of four things: an existing assignment to a named flight crew; be on the existing list of candidates available for selection to a crew; have all the right qualifications to get selected as an astronaut; or have a lot of money and pay for a 'space tourist' seat, which will still need a large set of requirements to be met and a lot of training before you are approved. If none of these categories fit, potential visitors need to get qualified and apply to become an astronaut or a cosmonaut through the national space agencies of European countries, Japan, Canada, the USA, or Russia.

To send experiments to the station is easier, but still requires researchers to conform to strict requirements and to apply through the same agency channels as those for astronauts. Experiments are divided into several different categories including microgravity, space environment, astrophysical, and Earth and atmosphere observations. Microgravity experiments are further divided up into different affiliations: life sciences, materials processing, and physics research.

Space environment studies involve

BELOW Anatoli Ivanishin gets a close haircut on the ISS, loose particles being vacuumed out of the air or trapped in the air filtration system. *(NASA)*

observations and measurement of the physical nature of space beyond the atmosphere of the Earth, and some experiments in this field of research are left exposed to the space environment for various periods before being returned to the inside of the ISS or brought back to Earth. Astrophysics involves studying the structure and nature of the physical universe and this can be conducted either by using data transmitted to Earth from experiments already on board or by providing new experiments. Observations of the Earth and the atmosphere usually involve digital images and data freely available to researchers.

A great deal of work has been conducted on microgravity experiments since the early 1970s and both Russia and America have carried out extensive research into new materials and pharmaceutical products promised by this early work. In some cases it has been found that techniques worked out for use on the ground are better and more efficient than those conducted in space while in other cases much has been learned about the behaviour of materials in a near-weightless environment.

Keeping it going

With the Shuttle retired, the only means of people getting to and from the ISS is by Russian Soyuz launched from Baikonur, but eventually there will be spacecraft developed by commercial companies in the US to carry crew and cargo back and forth. Already US cargo modules are being developed by commercial companies to take over the job done previously by Europe's ATV and Japan's HTV and these could be operational by the end of 2012. By 2017 it is hoped that people too will fly in commercially developed crew modules launched once again from the United States. Europe is committed to the ISS up to 2014 and beyond that it is hoping to remain

BELOW NASA's ISS Mission Control room at the Johnson Space Center, Houston, Texas, where operations have been conducted around the clock since 1998. *(NASA)*

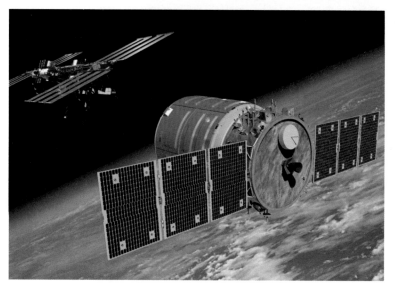

ABOVE Russia has its own ISS Mission Control outside Moscow and here operations with the Russian segment of the station are controlled. *(NASA)*

RIGHT Orbital Sciences is developing its Cygnus module tasked with leasing services to NASA for flying cargo to the ISS. *(NASA)*

ABOVE A private US company, SpaceX is developing a Dragon cargo freighter for delivering supplies to the station. *(SpaceX)*

a part of the programme to the end of ISS operations, presently scheduled for 2020, by providing add-on services to the space station itself or to the next generation of NASA manned flights.

Beyond 2020 the future is less clear but the ISS may play a role in helping develop the next goal for US human space flight into deep space, to places beyond Earth orbit. NASA is developing a new manned spacecraft called the Orion Multi-Purpose Crew Vehicle (MPCV) and a new rocket called the Space Launch System (SLS). With this combination NASA hopes to conduct missions to the Moon, the asteroids, and perhaps to Mars, but not until at least 2025 when Orion and the SLS will have been tested on previous flights. It is likely that the ISS will continue in operation into the third decade of this century. It has much to give in terms of scientific research opportunities in microgravity and in the biomedical information it can provide for astronauts and cosmonauts embarking from Earth on voyages deep into the solar system.

LEFT Modules such as Orbital Sciences' Cygnus are expected to form the basis for the development of crew-carrying commercial spacecraft by 2017. *(Orbital Sciences)*

Glossary

ACBM Active Common Berthing Mechanism
ACES Atomic Clock Ensemble in Space
AM Avionics Module
AMS Alpha Magnetic Spectrometer
APAS Androgynous Peripheral Assembly System
APCU Assembly Power Converter Unit
APM Attached Pressurised Module
ATA Ammonia Tank Assembly
ATCS Active Thermal Control System
ATV Automated Transfer Vehicle
BCDU Battery Charge/Discharge Unit
BGA Beta Gimbal Array
Canadarm2 (see SSPDM)
CBM Common Berthing Mechanism
CETA Crew & Equipment Translation Aid
CL Crew Lock
CMG Control Moment Gyroscope
CNES Centre National d'Études Spatiale (French Space Agency)
COLBERT Combined Operational Load-Baring Resistance Treadmill
CTC Cargo Transportation Container
DARPA Defense Advanced Research Projects Agency
DC Docking Compartment (Pirs)
DCSU Direct Current Switching Unit
DDCU DC-to-DC Converter Unit
Dextre (see SSRMS)
DM Docking Module
DMS Data Management System
EADS European Aeroneutronic Defence and Space Company
ECLSS Environmental Control and Life Support System
EEACTS Early External Active Thermal Control System
EFBM Exposed Facility Berthing Mechanism
EF Exposed Facility
EL Equipment Lock
ELM-ES Experiment Logistics Module-Exposed Section
ELM-PS Experiment Logistics Module-Pressurised Section
EMU Extravehicular Mobility Unit
EOTP Enhanced Orbital Replacement Unit Temporary Platform
ESA European Space Agency
ESP External Stowage Platform
ESTEC European Space Research & Technology Centre
ETCS External Thermal Control System
EuTEF European Technology Exposure Facility
EVA Extravehicular Activity

EWIS External Wireless Instrumentation System
ExPRESS Expedite Processing of Experiments to the Space Station
FGB Functional Cargo Block (acronym in Russian)
FIR Fluids Integrated Rack
FRAM Flight Releasable Attachment Mechanism
GMT Greenwich Mean Time (also known as Universal Time or UTC)
GPS Global Positioning System
HPGT High Pressure Gas Tank
HRF Human Research Facility
HTV H-II Transfer Vehicle
IATCS Internal Active Thermal Control System
ICC Integrated Cargo Carrier
ICC-VLD Integrated Cargo Carrier-Very Light Deploy
IEA Integrated Electronics Assembly
ISPR International Standard Payload Rack
ISS International Space Station
ITS Integrated Truss Segment
JEM Japanese Experiment Module
JEMRMS Japanese Experiment Module Remote Manipulator System
KSC Kennedy Space Center
LCA Lab Cradle Assembly
LDU Linear Drive Unit
LMC Light Multi-purpose Carrier
LSM Logistics Single Module
MARES Muscle Atrophy Research & Exercise System
MBS Mobile (Remote Servicer) Base System
MELFI Minus-80° Laboratory Freezer
MISSE Material (for) International Space Station Experiments
MPCV Multi-Purpose Crew Vehicle
MPLM Multi-Purpose Logistics Module
MRM Multi-purpose Research Module
MSFC Marshall Space Flight Center
MSRR Materials Science Research Rack
MT Mobile Transporter
MTFF Man-Tended Free Flyer
MTSAS Module-to-Truss Segment Attachment System
N₂O₄ Nitrogen tetroxide
NASA National Aeronautics and Space Administration
Nm Nautical mile
OBSS Orbiter Boom Sensor System
OGS Oxygen Generation System
OMV Orbital Manoeuvring Vehicle
ORU Orbital Replacement Unit

OTV Orbital Transfer Vehicle
PCBM Passive Common Berthing Mechanism
PCU Power Control Unit
PDGF Power Data Grapple Fixture
PGF Power Grapple Fixture
PLC Pressurised Logistics Carrier
PLM Permanent Logistics Module (rebranded from Leonardo MPLM)
PLM Pressurised Laboratory Module
PM Propulsion Module
PMA Pressurised Mating Adapter
PMA Pump Module Assembly
POA Payload Orbital Replacement Unit Accommodation
PPF Polar Platform Facility
PTU Power Transfer Unit
PV Photo Voltaic
PVAA Photo Voltaic Array Assembly
PVM Photo-Voltaic Module
PVR Photo-Voltaic Radiator
RGA Rate Gyro Assembly
RMS Remote Manipulator System
RRM Robotic Refuelling Mission
RSP Resupply Stowage Platform
RSR Resupply Stowage Rack
RSU Roller Suspension Units
RWS Remote Work Station
SABB Solar Array Blanket Box
SAFER Simplified Aid for EVA Rescue
SAGANT Space-to-Ground Antenna
SARJ Solar Array Rotary Joint
SASA S-band Antenna Support Assembly
SAW Solar Array Wing
SFOG Solid Fuel Oxygen Generator
SLS Space Launch System
SPDM Special Purpose Dextrous Manipulator (Detre)
SSPTS Shuttle-Station Power Transfer System
SSRMS Space Station Remote Manipulator System
SSU Sequential Shunt Unit
STS Space Transportation System
TDRSS racking & Data Relay Satellite System
TRRJ Thermal Radiator Rotary Joint
TVIS Treadmill & Vibration Isolation System
UCCAS Unpressurised Cargo Carrier Attach System
UDMH Unsymmetrical Dimethyl Hydrazine
UPLC Unpressurised Logistics Carrier
VLD Vertical Light Deployable
WORF Window Observational Research Facility
ZGSR Zero-G Stowage Rack

Index